D0706392

EVOLUTION AND THE
FOUNDERS OF PRAGMATISM

Evolution and the Founders of Pragmatism

PHILIP P. WIENER

**WITH A FOREWORD
BY JOHN DEWEY**

WITHDRAWN

THEODORE LOWNIK LIBRARY
BENEDICTINE UNIVERSITY
5700 COLLEGE ROAD
LISLE, IL 60532-0900

**University of Pennsylvania Press
Philadelphia**

144.3
w647e
1972

TO
ARTHUR O. LOVEJOY

Copyright © 1972 by the University of Pennsylvania Press, Inc.

This book was originally published in 1949 by Harvard University Press

First *Pennsylvania Paperback* edition 1972

ISBN: 0-8122-1043-3

Printed in the United States of America

Preface

This study is based on a series of investigations into the genesis of a brood of philosophical doctrines loosely labeled "American pragmatism." Those who judge the merits of pragmatism should be clear which of its varieties they are criticizing, who held the complex ideas being judged and in what context. This historical study aims to shed light on the matrix of ideas about evolution from which diverse meanings of pragmatism emerged in the thinking of a brilliant group of philosophical liberals who met around Harvard in the 1860's and early 1870's.

Charles Darwin is reported to have observed that there were enough brilliant minds at the American Cambridge in the 1860's to furnish all the universities of England. The Harvard group considered in this book as the founders of the many-sided doctrine of pragmatism are reported by one of them to have met informally about 1870 in a Metaphysical Club. There is no doubt that they were greatly impressed with the far-reaching implications of the Darwinian controversy. The writings of the members of this distinguished group—Chauncey Wright, Charles S. Peirce, William James, John Fiske, Nicholas St. John Green, and Oliver Wendell Holmes, Jr.—clearly reflect in their different fields the whirlpool effects of evolutionism and its diverse interpretations. They elaborated, in their own distinctive manner, the philosophical significance of evolution for the studies in which they later severally distinguished themselves. The resulting pragmatisms were as diverse as their personal interests and attainments, but they shared a common faith in man's fallible intelligence to meet coöperatively the many temporal contingencies on which his cultural survival depends. There is scarcely any basic problem in the natural and social sciences to which this group did not make some important contribution to our thinking. The pragmatic legacy inherited by twentieth-century American thought does not provide a neat, finished system, legislating for intellectual or moral questions, but it does provide a philosophical bulwark in the defense of freedom of inquiry and experimental ways of thinking.

Philip P. Wiener

Acknowledgments

I hesitate to acknowledge the generous aid of so many good minds who may frown at the result, but hasten to absolve them of any responsibility for any of the inadequacies of this book.

For my initial introduction to the vast historical theme of the idea of evolutionism and to the problem of tracing its appearance in the birth of diverse pragmatisms, I am deeply indebted to the incisive learning and friendly encouragement of Arthur O. Lovejoy. My earlier teachers, Morris R. Cohen and John Dewey, gave me my first insight into pragmatism as both a reflection and critique of American culture.

For suggesting a search for new material on Chauncey Wright I am indebted to Herbert W. Schneider; Mrs. Dorothy Pearson Abbott and Mrs. Chauncey Wright Pearson, both at Smith College, were kind enough to give me complete access to the personal papers and correspondence of Chauncey Wright. Clarence I. Lewis was very helpful in guiding me among the Peirce manuscripts at Widener Library. I owe very much in my chapter on William James to Ralph Barton Perry's great work and to Dickinson Miller's personal insight. Frederick Green kindly put at my disposal much that has not been heretofore recorded of his father's life and legal philosophy. Mark DeWolfe Howe of the Harvard University Law School has been most generous in furnishing me with material and helpful criticisms concerning Green and Holmes; indeed, without his friendly encouragement this book would not have appeared. Langdon Warner has supplied me with biographical data about his father, Joseph B. Warner.

I am also very grateful to those who have read various portions of the manuscript and suggested rewriting earlier drafts. John H. Randall, Jr. has aided me in the first versions of some parts published in the *Journal of the History of Ideas*. These parts have been considerably revised and supplemented with new material. Revisions of some chapters were suggested after they were read by my friends Jerome Rosenthal and Nathan Berall. Jerome Hall was kind enough to send me criticisms of the chapter on Nicholas St. John Green. Perry Miller offered valuable suggestions and much encouragement. To Phoebe deKay Donald of the Harvard University Press I am more than grateful for her skillful editorial labors in putting my sprawling manuscript into the shape of a book that may be useful to students of American thought.

Without the aid of research grants from the American Philosophical Society I could not have gathered the materials for this study. The Research Committee and administrative officers of the City College of New York have aided me by a reduction in my teaching schedule.

Little, Brown and Co. have permitted me to quote from Ralph Barton Perry's *The Thought and Character of William James* (Boston, 1935), and Charles Scribner's Sons have been similarly courteous in permitting me to quote from Henry James's description of Chauncey Wright in *Notes of a Son and Brother* (New York, 1914, 1942). Mrs. Ethel Otis D. Fisk has kindly given me permission to quote from her edition of *The Letters of John Fiske* (New York: Macmillan Co., 1940) and from John Spencer Clark's *The Life and Letters of John Fiske* (Boston: Houghton Mifflin Co., 1917).

There is no adequate way of expressing here what I owe to the love and patience of my wife Gertrude.

P. W.

Contents

Foreword

There is an old English saying that good wine needs no bush. Dr. Wiener has provided for all who are interested in the development of the intellectual life of North America a generous supply of cultural wine, a wine with body and flavor. Being asked to write a brief Foreword, I gladly complied; not because the chapters forming the text are in need of any praise or recommendations, but because it is an honor to have even a remote connection with an authoritative study of the origin and early development of the intellectual and moral ferment generated in this country by the new scientific developments of the middle of the nineteenth century. For the work is more, very much more, than a study of the initiation of a particular philosophical Ism, and is also very much more than a well-documented and thoroughly informed presentation of material that completely annihilates the misstatements about the aim and tenor of the founders of the pragmatic movement that have flourished. Argument to refute the misrepresentations that were widely current a generation ago (and that are still put forth from some quarters) is not engaged in; Dr. Wiener's thorough familiarity with the facts, and his extraordinarily well-documented and penetrating recital of them, leave the misconceptions in the void of ignorance—not to say willful stupidity—in which they originated.

A British critic of the later period of the movement stated that pragmatism was an expression and organized reflection of American commercialism. Although years earlier he was not obsessed with anti-American prejudice, he had come within speaking distance of the movement by saying it was based on the inductive phase of modern science. But anyone who reads the first chapter of Philip Wiener's book will see how superficial is even that approach to understanding. In the last few years the work of Peirce has been receiving an attention long overdue. But if there is any account of the context which provided the soil, light, and atmosphere of his vital contribution that begins to compare with that of Dr. Wiener, it has missed my notice. One realizes the ineptitude of the method of pigeonholing classification of philosophical writings when one compares it with the method of placing them in the setting of a new and vital movement in culture which extends far beyond the confines of technical philosophy. This latter contribution Dr. Wiener makes; he has not merely contributed to the understanding of a significant distinctively American philosophical movement, but he has also provided a formation and exemplification of an enlightened and liberal method for dealing with the philosophical activity of *any* historical period.

His treatment of the "Foundations of Pragmatism" in terms of a deep, moving cultural current, takes us to a time when America was still a symbol of the dawn of a better day and was full of hope infused with courage. It will be a happy day for American philosophy if, after a period of loss of nerve, this stirring account of the initial period of a *movement* (not an Ism) recalls the wandering thoughts of American teachers of philosophy to the creative movement to which they belong as Americans, whatever school they belong to professionally.

John Dewey

EVOLUTION AND THE
FOUNDERS OF PRAGMATISM

"The candle that is set up in us

shines bright enough for all our purposes."

JOHN LOCKE

CHAPTER ONE

The Background of Science
and Natural Theology

1. SCIENTIFIC IDEAS OF THE MID-NINETEENTH CENTURY

Philosophy, Bertrand Russell finds in his *History of Western Philosophy*, has been, and still is, something intermediate between theology and science. During the third quarter of the nineteenth century American philosophical theology was profoundly stirred by the scientific ferment and technological progress of the times. Though natural theology now belongs in the museum of fossilized ideas, it was at that time very much alive in the minds of scientists and educational authorities, reared on theology but anxious to adjust their ideas to new scientific discoveries. Darwinism and pragmatism were able to combat their conservative theological adversaries only because of the powerful impetus of scientific advances in the second half of the nineteenth century. While religious sects continued to war on one another, while economic strife grew more and more intense between labor and capital, farmers and manufacturers, rural and urban interests, West and East, North and South, any philosophical student not engrossed in these battles could look with relief and hope, as the founders of pragmatism did, to the steady and brilliant discoveries by scientific men of different religious, political, national, and racial origins. The infinite perspective of science was displacing the transcendental brooding of theology, but not without a clash of ideas. Let us dwell briefly on some of the scientific and theological ideas in the intellectual atmosphere breathed by the founders of pragmatism.

They knew, for example, of the remarkable discovery of the planet Neptune in 1846, resulting from the mathematical calculations of the Frenchman Ulpian J. J. Leverrier and the Englishman John Couch Adams, confirmed by the telescopic observations of a German astronomer, Galle. Peirce and Wright were practicing astronomers in their twenties, and were surely impressed by the new jewels worn by the disinterested queen of the sciences. They were philosophical physicists in a glorious era of physical discoveries. In 1849 and 1850, Fizeau and Foucault measured the velocity of light in air, and also showed, against Newton's authority, that its speed diminished in denser media. Joule had in 1847 experimentally determined the mechanical equivalent of heat, whose properties had long eluded precise formulation, though much had been discovered of its gross properties by earlier experimenters like Black (1728–1799), Rumford (1753–1814), and Sadi Carnot (in his sole work, *Réflexions sur la puissance motrice du feu*, 1824). The greatest physical generalizations of the midnineteenth century were the kinetic theory of gases and the laws of thermodynamics (conservation of energy and law of entropy). The kinetic theory of gases contains the very significant idea of a statistical mode of accounting for the molecular laws of diffusion and viscosity, and for unifying the laws of Boyle, Charles, and Avogadro. This was done through statistical calculations of the average or probable net behavior of large numbers of individual particles. There had been an earlier use made of statistics in the seventeenth-century studies of mortality tables and insurance rates (John Graunt, Sir William Petty, Jan De Witt); Leibnitz, Pascal, Laplace, Demoivre, the Bernouillis, and other mathematicians had worked out the rules for calculating probabilities; Cournot, Quetelet, and Buckle applied the statistical method to social questions. Though Leibnitz and Hume had shown the philosophical importance of probability in empirical matters, it was not until Boole, De Morgan, and Peirce mathematically overhauled traditional formal logic that the logic of probability was put on a more scientifically useful basis.

Now what has all this history of physics and of the logic of chance in physical and social science to do with evolution and pragmatism? A rather full answer (albeit in metaphysical shape) is furnished us by one of the founders of pragmatism, Charles Peirce, himself a physicist, logician, and historian of science:

Natural selection, as conceived by Darwin, is a mode of evolution in which the only positive agent of change in the whole passage from moner to man is fortuitous variation. To secure advance in a definite direction chance has to be seconded by some action that shall hinder the propagation of some varieties or stimulate that of others. In natural selection, strictly so called, it is the crowding out of the weak. In sexual selection, it is the attraction of beauty.

The *Origin of Species* was published toward the end of the year 1859. The preceding years since 1846 had been one of the most productive seasons—or if extended so as to cover the great book we are considering, *the* most productive period of equal length in the entire history of science from its beginnings until now. The idea that chance begets order, which is one of the corner-stones of modern physics . . . was at that time put in its clearest light. Quetelet had opened the discussion by his *Letters on the Application of Probabilities to the Moral and Political Sciences,* a work which deeply impressed the best minds of that day, and to which Sir John Herschel had drawn attention in Great Britain. In 1857, the first volume of Buckle's *History of Civilization* had created a tremendous sensation, owing to the use he made of the same idea. Meantime, the "statistical method" had, under that very name, been applied with brilliant success to molecular physics. Dr. John Herapath, an English chemist, had in 1847 outlined the kinetical theory of gases in his *Mathematical Physics;* and the interest the theory excited had been refreshed in 1856 by notable memoirs by Clausius and Krönig. In the very summer preceding Darwin's publication, Maxwell had read before the British Association the first and most important of his researches on this subject. The consequence was that the idea that fortuitous events may result in a physical law, and further that this is the way in which those laws which appear to conflict with the principle of the conservation of energy are to be explained, had taken a strong hold upon the minds of all who were abreast of the leaders of thought. By such minds, it was inevitable that the *Origin of Species,* whose teaching was simply the application of the same principle to the explanation of another "non-conservative" action, that of organic development, should be hailed and welcomed. The sublime discovery of the conservation of energy by Helmholtz in 1847, and that of the mechanical theory of heat by Clausius and by Rankine, independently, in 1850, had decidedly overawed all those who might have been inclined to sneer at physical science. Thereafter a belated poet still harping upon "science peddling with the names of things" would fail of his effect. Mechanism was now known to be all, or very nearly so. All this time, utilitarianism—that improved substitute for the Gospel—was in its fullest feather; and was a natural ally of an individualistic theory . . . there were several elements of the Darwinian theory which were sure to charm the followers of Mill. Another thing: anaesthetics had been in use for thirteen years. Already, people's acquaintance with suffering had dropped off very much; and as a consequence, that unlovely hardness, by which our times are so contrasted with those that immediately preceded them, had already set in, and inclined people to relish a ruthless theory. The reader would quite mistake the drift of what I am saying if he were to understand me as wishing to suggest that any of those things (except per-

haps Malthus) influenced Darwin himself. What I mean is that . . . the
extraordinarily favorable reception it met with was plainly owing, in large
measure, to its ideas being those toward which the age was favorably disposed,[1] especially because of the encouragement it gave to the greed-
philosophy.[2]

We shall defer to later pages discussion of the divergent opinions
among the founders of pragmatism concerning the ethical implications of Darwin's work. We shall find Justice Holmes recalling
that "it was the influence of the scientific way of looking at the
world" (Darwin, Spencer, Buckle)—"it was in the air"—which
encouraged his own youthful skepticism of theological explanations, apart from his falling in with the abolitionists, "some or
many of whom were sceptics as well as dogmatists." * Holmes followed Wright rather than Peirce on the question of the meaning
of chance and its relation to metaphysical necessity and to practical knowledge, but these again are topics of later pages. Let us
return to the scientific use made of the idea of chance just about
when Darwin's work appeared. We note that the greatest American physicist then, Josiah Willard Gibbs,[3] was soon to obtain
Maxwell's highest praise for his work on statistical mechanics
and the phase-rule. Maxwell called the calculus of probabilities
"Mathematics for practical men," as J. T. Merz records in his *History of European Thought in the Nineteenth Century*.[4] Merz,
however, fails to note that Maxwell subordinated the "practical
man's" statistical view to the theoretical ideal of mathematical
dynamics. In preferring the "dynamical" to the "statistical" type
of theory, Maxwell inherited the rationalistic ideal of the complete determination by fixed law of the behavior of each individual particle of the universe. In America, Gibbs and Peirce saw
the far-reaching consequences of the statistical view of laws more
clearly than the greatest of contemporary physicists abroad (excepting Boltzmann). In path-breaking mathematical and logical
analyses, respectively, they showed that physical laws would have
to abandon the classical rationalistic ideal of certainty concerning
individual behavior and seek for order in the general average behavior of groups.

In biology the statistical view of nature meant, to a careful
investigator like Darwin, that *the order of species emerges from*

* See p. 173.

the net result of minute variations of individual organisms subject to variable environmental conditions. There is an orderly selection by nature of cumulative adaptations open to observation, though each individual variation, preserved or eliminated by the environment, may not be initially accounted for or predicted. Such was the tremendously sweeping character of the idea of chance variations, appearing in Quetelet's and Buckle's sociological statistics,[5] invading exact physical science, and supporting the revolutionary hypothesis of Darwin concerning the origin of species.

No wonder, then, that all the founders of pragmatism in one way or another were affected by the idea of the contingency of nature and the irreducible individuality of man. We shall recognize the idea in Chauncey Wright's "cosmic weather," in Peirce's cosmological tychism, in James's attack on the "block universe" and in his Carlylean theory of great men in history, in Holmes's "bet-abilitarianism," and in Nicholas St. John Green's repudiation of the scholastic chain of preordained causation. Of course, the idea of contingency became transmuted in the diverse contexts of the physical and human fields cultivated by the founders of pragmatism. But that is the pragmatic fate of all ideas, namely, to grow in meaning with different applications. The naturalistic poet Lucretius, with his inspired vision of the evolution of the world by chance combinations of invisible atoms, would have been delighted to learn what a tremendous significance the swerve of the atom could have in displacing the older gods, but he would not have understood a word of the statistical language of modern physical, biological, and social science.

The historian of ideas has to unite the minutiae of historical research with logical analyses of the structure and implications of ideas and of their modifications in diverse contexts. The idea of evolution meant different things to pre-Darwinian and post-Darwinian scientists and philosophers, though there are lines of continuity also. We find the idea of evolution in the ancient Greek scientific philosophers and in the poet-philosopher, Lucretius, who speculated on the emergence of the fixed seeds of life from the slime of matter. In Aristotle's zoological philosophy it meant, among many other things, the internal development of a preformed embryo toward an adult type. The types fell within a fixed hierarchy of species. To use Professor Lovejoy's apt terms, the neo-Platonists bequeathed the ideas of a gradation and plenitude of fixed species to cosmology and biology. Aristotle's teleological idea that nature does nothing in vain—a phrase which occurs in his *Politics*—failed in the eyes of modern scientists to establish an objective continuity among the links of his version of the scale of being. The Scholastics fused these semiscientific, semispeculative ideas with their own unverifiable idea of a divine providence. How later theologians and metaphysicians accommodated their ideas of creation to their conceptions of man's place in a closed, determined universe is a long story. That story of the persistence of the Scholastic view of special creation down through the nineteenth century, even after Darwin's work appeared, helps us understand the resistance to his ideas by theologically minded and respected scientists like Edward Hitchcock, Paul Chadbourne, Louis Agassiz, and St. George Mivart.

Even late in the nineteenth century, scientists, especially in biology and social sciences, were expected to show how their findings were compatible with older religious and ethical systems, despite Kant's philosophical separation of science from ethics and religion. Kant's philosophy was daily discussed by Chauncey Wright and Charles S. Peirce during the first years of the Darwinian controversy. Wright argued for a Kantian and nonmetaphysical separation of scientific theories like Darwin's from value judgments, while Peirce tried to find a higher metaphysical unity of scientific and religious notions in the post-Kantian fashion of Schelling. An important distinction arises here between *evolution* as a scientific theory in a specific field and *evolutionism* as a generalization invading every field of study from biology and cosmology to sociology and philosophy of history. This distinction, made out so clearly in Professor Lovejoy's larger historical studies of the pre-Darwinian ideas of evolution and post-Darwinian evolutionisms, will be very useful for our more restricted study of the relationship of nineteenth-century evolutionism to the origins of pragmatism. We shall see that one of the chief questions discussed by the founders of pragmatism, beginning with Chauncey Wright, was how far one could legitimately apply Darwin's hypothesis of natural selection to subjects other than biology.

Wright's deep interest in the history, methods, and philosophy of physical and biological sciences led him on more than one occasion to point out to his contemporaries the important intellectual changes which the idea of evolution was bringing about. In an obituary notice of the great geologist, Sir Charles Lyell, Wright, in his own last year, indicated the historical and philosophical significance of Lyell's late adoption of the Darwinian theory of "development," a then current synonym for "evolution."

At the end of the last century, the transmutation or development theory was independently and almost simultaneously proposed by three great thinkers, Goethe, Geoffroy St. Hilaire, and Erasmus Darwin. Its final triumph in our own day was almost a direct consequence of the principles adopted by Lyell from Hutton and the Huttonians, and urged so clearly and effectively by him in his *Principles*. Yet Lyell—and this was an interesting exhibition of a worthy trait of his mind—resisted the theory of development for a long time; until after the publication of that most remarkable book of the century, Darwin's *Origin of Species* . . . This change [in the tenth edition of Lyell's *Principles*] gives his masterpiece a greater logical completeness and coherency than it had ever had before, and redounded to his credit in this way quite as much as in the exhibition it gave of his openness of mind to scientific arguments, or of the moderation of the conservatism which characterized him as a true English Liberal.

Hutton, the knowledge and practice of whose principles Lyell did so much to extend, was the first to declare that geology was in no wise concerned with "questions as to the origin of things." By "origin" was then meant the origin of natures which things have and their first introduction in the theatre of the world. That cosmology should have been so far banished in half a century from zoological conceptions that Mr. Darwin could use, without incurring serious misunderstanding, the title *Origin of Species* for his great work, is evidence of the progress made . . . "Origin" now means how things go from one determinate appearance to another . . . The word "species" now means not an absolute, but only a comparative fixity of character, [so that in Darwin] the two terms appear with modern non-scholastic meanings . . .

There appears to be a strong natural association of religious feeling with the idea of stability; and three wrongly consecrated stabilities—that of the earth, that of its continents, and that of its forms of life—have one after the other given way to the progress of knowledge, and though with obstinate resistance from religious sentiment, without permanent injury to religion . . .

Lyell had the qualities in mind and character of the highest modern social and scientific culture.[6]

Thomas H. Huxley also pointed out that Sir Charles Lyell "was the chief agent in smoothing the road for Darwin. For consistent uniformitarianism postulates evolution as much in the organic as in the inorganic world." [7] The history of pre-Darwinian evolutionism falls beyond the scope of this book. We are here concerned with the more proximate background or matrix of scientific and philosophic ideas that help us understand the scientific success of Darwin's theory of chance variations and the mechanism of natural selection, and these are more pertinent to the reorientation of philosophy inaugurated by the founders of pragmatism. I have referred to the great strides made through the use of observational and statistical, experimental and mathematical methods in physical, biological, and social sciences after 1845, resulting in the discovery of fundamental theories: the conservation of energy, the kinetic theory of gases, the second law of thermodynamics, the evolution of the earth's crust and of the fossils found therein, the stages of embryological development, the principles of domestic breeding, Quetelet's, Comte's, and Buckle's sociologic generalizations, Tylor's laws of the development of primitive societies, Maine's theory of the passage from status to contract, and the Malthusian law of population growth (which Darwin said suggested to him the idea of the struggle for survival).

In his defense of Darwin, Chauncey Wright underscored the statistical character of biological laws, even if the governing averages suggested more ultimate laws of "real tendencies." What was more important to him than the "ultimate" metaphysical status of such laws was their independence of theology and their relevance to the true grounds of social institutions, law, and morality:

Nearly all laws of heredity are properly laws of averages, against which, of course, evidences may be massed by a partial or one-sided induction. They can in general be held to be true only as governing averages; or at least as being laws of such real tendencies as mask and modify and even countervail one another . . . The power of Education is, indeed, the scientific foundation of law and moral responsibility; but heredity is also great, and even adds to, instead of diminishing, the true grounds of social institutions, the rights of law and punishment . . . Intelligence and strength in union or brotherhood have long been the vantage grounds of the human contest in the competition of race with race, or tribe with tribe, and in opposition to unfavorable external conditions. And biology affords a proof, in the natural genealogical tree, of the merely natural brotherhood of man, quite as impressive as the theological derivation of the race from a single pair; but does not make so complete and emphatic a reference back to one concentrated responsibility for human weaknesses and sins.[8]

It was in line with his pluralistic empiricism that Wright raised critical objections to the extension of Darwin's theory of natural selection beyond biological and psychological phenomena to cosmological and theological domains. Before and after Darwin's *Origin of Species* the thinking of natural philosophers was dominated by vague unifying principles of vitalism and teleology, for natural theology was firmly established as the proper framework of the sciences.

In order to illustrate the metaphysical and theological climate of scientific opinion in the United States during the third quarter of the nineteenth century, we shall examine two typical texts, one by the highly respected geologist Edward Hitchcock, the other, a post-Darwinian work by the popular Lowell lecturer on psychology, Paul Chadbourne. Specimens of their scientifically inadequate but widespread views need only to be quoted in order to understand why Wright and his friends sought more rigorous analyses of the scientific and moral questions uncritically lumped together by the natural theologians of their day.

2. A RELIGIOUS GEOLOGIST'S ORGANIC VIEW OF NATURE

Let us first look into "The Telegraphic System of the Universe" erected in the twelfth lecture of *The Religion of Geology and Its Connected Sciences* "by the learned President of Amherst and Professor of Geology and Natural Theology, Edward Hitchcock." His inspired book reached its eleventh thousand printing in Boston by 1856. Note in the following passage how he assimilates the principles of mechanics to the idea of an organic chain of vital influences, and how he then proceeds to the electrification of the whole golden chain by means of an elaborate wiring system reaching every creature throughout space and time:

The records of zoology and botany afford endless illustration of this subject [the equilibrium of organic nature]. But the great truth which they all teach is, that so intimately are we related to other beings, that almost every action of ours reacts upon them for good or evil; for good, upon the whole, when we conform to the laws which God has established, and for evil, when by their violation we disturb the equilibrium of organized nature, and produce irregular action. In this latter case, we cannot tell where the disturbance, thus introduced, will end; for it is not a periodical oscillation, like the perturbations of the heavenly bodies, nor a mere change of position and intensity by mechanical forces.

But does not this law of mutual influence between organic beings extend to other worlds? Why should it not be transmitted by means of the luminiferous ether to the limits of the universe? Who knows but a blow struck upon a *single link of organic beings* [my italics] here may be felt through the whole circle of animate existence in all worlds? That is a narrow view of God's work, which isolates the organic races of this globe from the rest of the universe. The more philosophical view throws the *golden chain of influence* [my italics] around the whole animal creation, whether small or great, near or remote.[9]

The electrification of the golden chain of organic beings was accomplished by Hitchcock as follows:

What a centre of influence does man occupy! It is just as if the universe were a tremulous mass of jelly which every movement of his made to vibrate from the centre to the circumference. It is as if the universe were one vast picture gallery, in some part of which the entire history of the world, and of each individual, is shown on canvas, sketched by countless artists, with unerring skill. It is as if each man had his foot upon the point where ten thousand telegraph wires meet from every part of the universe, and he were able, with each volition, to send abroad an influence along these wires, so as to reach every created being in heaven and on earth. It is as if we had the more than Gorgon power of transmuting every object around us into forms beautiful or hideous, and of sending that transmuting process forward through time and through eternity. It is as if we were linked to every created being by a golden chain, and every pulsation of our heart or movement of our mind modified the pulsation of every other heart and the movements of every other intellect. Wonderful, wonderful is the position man occupies, and the part he acts! And yet it is not a dream but the deliberate conclusion of true science.[10]

One of the effects of evolutionism, even before Darwin, was to make the human mind more time-conscious than it had ever been under the reign of Euclidean and Newtonian science. Buffon had been bold enough in the eighteenth century to extend the age of man's life on earth from six to sixty thousand years;

but soon after, the geologic and paleontological investigations of Lyell and Cuvier implied much longer stretches of time for the explanation of the variety of extinct as well as of living species and their habits. The greatest defender in scientific Germany of Darwin's views, Ernst Haeckel, maintained that biologic evolution requires a duration of time which the human intellect cannot imagine, just as astronomy since Römer's measurement of the velocity of light has introduced unimaginable distances.

> In the same way as the distances between the different planetary systems are not calculated by miles but by Sirius-distances, each of which comprises millions of miles, so the organic history of the earth must not be calculated by thousands of years, but by paleontological and geological periods, each of which comprises many thousands of years, and perhaps millions, or even milliards of thousands of years.[11]

Haeckel needed all the time possible in order to give his Lamarckian factors of use and disuse ample opportunity to effect modifications of species into new species. Instincts he regarded as "habits of the soul acquired by adaptation and transmitted and fixed by inheritance through many generations."[12] In any case, this enormous extension of time led Haeckel to point out the unsuitability of our ordinary units of time-measurement and the relativity of scientific units of time. The human individual's life-span "is not suitable as a standard for the measurement of geologic periods." Other organisms have life-spans ranging from a few hours to five thousand years. "This comparison brings the relative nature of all measurement of time clearly before us."[13]

Now relativity and temporalism do not fit into the Kantian view that time is an a priori form of sensory intuition and that scientific knowledge is absolutely conditioned by eternal categories.

Hitchcock was fully aware that, as astronomy, especially since the time of the telescope, reveals the expansion of our ideas regarding the extent of the material universe, so "geology has given great enlargement to our knowledge of the divine plans and operations in the universe," for "it expands our ideas of the time in which the material universe has been in existence."[14] The microscope is "the sixth step in man's knowledge of Jehovah" and enables man to penetrate deeper into the lower "infinitesimal" regions of the chain of being, "teeming with countless millions even of organic beings, of a size much more diminutive than those yet discovered, and with inorganic beings too minute for the imagination to conceive." [15] Hitchcock's work thus illustrates the admixture of science, morals, and theology typical of pre-Darwinian thought about 1850. Even the metamorphic theory of Lyell is adopted in Hitchcock's estimation of the age of the earth as not contrary to Scriptures, "properly interpreted." The priority of life over nonliving matter is supported by the microscopic examination of the skeletal remains of silex, lime, and iron of marine animals and shells. The ancient adage, *Omnis calx e vermibus; omne ferrum e vermibus; omnis silex e vermibus,* is not so extravagant, thought Hitchcock, and he quoted Dr. Mantell's *Wonders of Geology* to support further this hylozoistic geology: "Probably there is not an atom of the solid materials of the globe which has not passed through the complex and wonderful laboratory of life."

Geology shows us that the present system of organic life on the globe is but one link of a temporal series finite at one end (to allow Creation), but reaching infinitely forward. "Revelation describes only the existing species, leaving to science the task and privilege of lifting the veil that hangs over the past, and to disclose other economies that have passed away." [16] The classifications of species of Agassiz and of Deshayes are accepted by Hitchcock as fitting into this providential scheme, which best permits us to read the thoughts of the Creator. Since that "immense remote time"—greater than a thousand times the six thousand years formerly accepted—when the material universe was created out of nothing, "matter has passed through a multitude of changes, and been the seat of numerous systems of organic life, unlike one another, yet all linked together into one great system by a most perfect unity; each minor system being most beautifully adapted to its place in the great chain, and yet each successive link becoming more and more perfect." [17]

So we have, only three years before Darwin's *Origin of Species,* a widely circulated and spirited expression of the idea of an organic temporalized chain with the links converted from species to individuals. Hitchcock's peculiar idea of the chain of being is dealt with in Professor Lovejoy's exposition of the continuity of nineteenth-century evolutionary ideas with eighteenth-century ideas of progress and with the more ancient neo-Platonic tradition of the golden chain of being. Of course, neither Darwin nor his contemporaries knew or needed to know anything of this philosophically intriguing story in order to play their role in its nineteenth-century development. The temporalizing of the chain of being in the Darwinian ideas of chance variations and natural selection by environmental action, implied the dissolution of its rigid, internal determinism and introduced an element of real novelty. Calvinist theologians like Hitchcock and McCosh, in order to avert the damaging effect of a head-on clash with the new scientific ideas, tried to reconcile them with their religious spiritualism. With this aim of fitting scientific truth into the "peculiar and higher truths disclosed by revelation," Hitchcock brought forward passages from the learned works of many theologians and scientists.[18]

Dr. W. M. Smallwood, in his *Natural History and the American Mind*, summarizes the situation of natural science in the United States up to the victory of Darwinism, when he notes succinctly that "the contributions of science were not organized as a consistent philosophy of nature until the period of Darwin." [19] The moralistic tendencies of American "natural philosophers" of that period are revealed in the lessons from nature found in children's books on natural history at that time, and in Paley's *Natural Theology* which "was widely used in both academy and college in the first half of the nineteenth century." [20] Hitchcock's *Elements of Geology* ran through thirty-one editions from 1840 to 1860.

3. A PSYCHOLOGIST'S POST-DARWINIAN NATURAL THEOLOGY

Our second text illustrates the climate of philosophical opinion among conservative natural theologians in the United States even after Darwin's rejection of fixed species and special creation had been defended by scientists like Asa Gray and his student Chauncey Wright, over the protests of Louis Agassiz.[21] William James referred to Chadbourne's theory of instinct as a typical example of the unscientific character of anti-Darwinian methods of psychologizing.*

* See Chapter V, sect. 2, below.

Natural theology was "shown" to be in harmony with the latest sciences by the Lowell lecturer of 1871, Paul A. Chadbourne, who published his lectures under the title, *Instinct: Its Office in the Animal Kingdom and Its Relation to the Higher Powers in Man.* His chief quarrel with natural selection is that it ignores the eternal truth that "the laws of human life and its conditions of progress are as fixed as the laws of gravitation and cohesion." [22] Chadbourne was the author also of *Relations of Natural History* and *Natural Theology.* In his Lowell lectures on instinct, which were undoubtedly known to the members of the Metaphysical Club, Chadbourne pleaded insistently for the cooperation of "Naturalists and Mental Philosophers in the full study of Man." [23] Only thus can the "social scientist" or "wise philanthropist" ameliorate the evils of society.

Another significant aspect of Chadbourne's work—recurring as a major topic of concern in the writings of all the founders of pragmatism—is a methodological awareness of the semantic difficulties of scientific and philosophical discussions of evolution. Wright and Peirce would have agreed fully with Chadbourne that when a writer introduces terms with new meanings, he should realize that the older connotations of these terms will persist in the minds of the reader and will interfere with the process of grasping the new meanings.[24]

How important language is in philosophy was indicated by Wright:

> The languages employed by philosophers are themselves lessons in ontology, and have, in their grammatical structures, implied conceptions and beliefs common to the philosopher and to the barbarian inventors of language, as well as other implications which the former takes pains to avoid. How much besides he ought to avoid, in the correction of conceptions erroneously derived from the forms of language, is a question always important to be considered in metaphysical inquiries.[25]

Wright and James, however, would have deeply disapproved of Chadbourne's philological conservatism in clinging to vague theological and metaphysical phrases like "the plan of creation" or "the instinct-like forethought which *is* the law of growth of plants and animals," [26] or "adaptation of means and ends such as justified itself to the Reason of man." [27]

Chadbourne found the old idea of "an organized whole" of predetermined purposes in nature a more satisfactory explanation of evolution than Natural Selection: "Within this one comprehensive plan, by which all beings seem to be related for their mutual good, we may consider the various subordinate plans for specific purposes." [28]

In this single cosmic plan, even well-adjusted individual creatures may be destroyed; hence, Chadbourne argues, the plan operates not by natural selection, but despite it.[29] He thus assumed with Agassiz[30] that Darwin's theory does not account for the elimination of the unfit, for it gives no metaphysical *reason* for the existence of favorable conditions in some cases and unfavorable ones in others. The simpler alternative principle of explanation is the vitalistic one which rests on our instinctive knowledge of the unity of life as determined by general Providence:

> A vital force or principle is so far uniform in its operations as to give us the simplest notion of life, which all have, although they may not be able to define it. And this principle that impresses us as *one*, under the name of *life*, manifests itself under hundreds of thousands of the most diverse forms of matter composed of the same elements, and takes for its cycle a single day as in the lower algae, or centuries, as in some of the higher animals. If asked now for the origin of this principle, or of its relationship to the great forces of nature, we are at present, as utterly at a loss to account for them as we are to account for gravitation itself or for the laws of its action . . . It is a characteristic of this principle in all its manifestations to demand and use as a means of putting forth its activities, the different elements and forces of the inorganic world. If asked for the origin of organized beings we come back in all our investigations where we want something given to begin the work with; as much so, as we need in Geometry axioms that cannot be demonstrated. When Mr. Huxley has carried us back to PROTOPLASM, we feel that we are as far off from the goal as ever.[31]

The logical alternative offered by Wright's nonmetaphysical view of this problem of mechanism and vitalism was to reject *all* empirically unverifiable principles about the ultimate constitution or origins of things. Chadbourne was aware of this positivistic view and was willing to adopt it as the method of approach in natural science, but declined to adopt it philosophically for the same reason he declined Huxley's agnosticism, namely, that both fail to supply the theological "bridge between the biological and physical worlds." [32]

Shall we then free ourselves of all preconceived notions of creation—of development, of Theology—of how things *ought to be*—or, at least leave them for future discussion and apply ourselves to the task of learning *what is*—in the department of nature which we propose to investigate? If we can do this, we shall gain for ourselves all the good which Positive Philosophy has ever had to offer as a guide in science, without committing ourselves to its dogmas. And this much should be said in favor of Positivism, that its method is the only true one for approaching natural science. Whether the human mind can stop, or ought to stop, within the limits which Positive Philosophy prescribes for it, is a very different question. [33]

That Chadbourne did not stop within any such positivistic limits is evident from the rest of his work; it contains such sweeping claims as that biological genera, families, and classes are as fixed as the chemical elements which only alchemists superstitiously believed they can transmute. [34] His metaphysical zeal came to a climax in an anthropocentric view of nature:

It is because all nature has, or may have relations to man,—because he is acted upon by every force, and related to all the changes of the organic and inorganic world,—that the study of nature is of any value. The whole physical universe is seen to centre in him . . . Man is the one point towards which all the rays of light in the physical universe seem to converge. [35]

Finally, we note that to Chadbourne's anti-Darwinian way of thinking, the concept of chance, so central in the physical, biological, and social sciences of his day, was of no logical value in any causal explanation. No intelligible phenomenon could ever result from the operation of chance, "even if the experiment could be tried every day for a geologic age. If it is said we ought to believe in such results, from the elements of variation and indefinite time, we cannot help it. We are satisfied that with the same data to rely upon, all men will not reach the same conclusions. It is possible that this may be partly the result of training, and it may arise from a constitutional difference among men in weighing proof."

This intrusion of psychology into the very logic of science, which Wright and Peirce condemned in their younger associate

William James, constituted a *reductio ad absurdum* of arguments like Chadbourne's. The scientific psychologist today would undoubtedly say that it well deserves its place of oblivion as scientifically unfit for survival in the struggle of ideas.

The historical importance of the founders of pragmatism consists in the fact that they challenged the validity of the theological fusion of scientific and ethical considerations by influential writers on "natural theology" like Hitchcock and Chadbourne. To face the contingency of nature and to render it intelligible without leaning on a providential intelligence was a daring philosophical experiment, and it was conceived in Peirce's Metaphysical Club which kept no minutes of its proceedings.

CHAPTER TWO

The Birthplace of Pragmatism:
Peirce's Metaphysical Club

Pragmatism poses some nice problems to both the historian of ideas and the philosopher. William James said it was "a new name for some old ways of thinking," [1] after having publicly credited to his friend Charles Peirce the authorship of the doctrine of "practicalism or pragmatism as he [Peirce] called it when I first heard him enunciate it at Cambridge in the early '70's." [2] But James was not a historian, and Peirce, in dire need of some public recognition at the turn of the century, [3] wrote sundry accounts of the genesis of pragmatism, some of them never published. In all these accounts Peirce constantly alludes to a Metaphysical Club, to the evolutionary controversy which had spread beyond biology, and to a paper which he had read to the Club setting forth a pragmatic theory of meaning and truth.

The most striking features of the Club are its informality, the rich diversity of interests and training of its members in natural and social sciences, logic, ethics, metaphysics, history, and legal practice. All of the members were or became lecturers at Harvard University at one time or another after 1870, when the new elective curriculum was introduced by the first nonclerical President of Harvard, the former professor of chemistry, Charles W. Eliot.

Certain incongruous features of Peirce's accounts of the Club and of the genesis of pragmatism justify Professor Perry's remark that the origins of pragmatism are obscure. First of all, the Metaphysical Club and its members are named in two accounts written by Peirce in 1905–06; the Club but not its members appears in an article published in 1908, and the members are mentioned without the name of the Club in an unpublished account. Secondly, in none of the writings, biographies, or letters of any of the Club's members have I come across any reference to the Metaphysical Club as named and described by Peirce. Thirdly, Peirce's recollections, going back more than thirty years, are not clear about the dates of his Club's meetings: "the early seventies," "in 1871," "in the sixties," and "after my return" [from Europe]. Why, then, did Peirce attach so much historical significance to the Club as the birthplace of pragmatism? How shall we relate the genesis of pragmatism to the evolutionary controversy? In this chapter I shall outline the answers to these questions, first by annotating Peirce's accounts of the Club, and then by sketching in a preliminary fashion the attitudes to the meanings of evolution held by the pragmatic members.

An early description of the Club, by Peirce, occurs in the following passage:

It was in the earliest seventies that a knot of us young men in Old Cambridge, calling ourselves, half-ironically, half defiantly, "The Metaphysical Club,"—for agnosticism was then riding its high horse, and was frowning superbly upon all metaphysics—used to meet, sometimes in my study, sometimes in that of William James.[4] It may be that some of our old-time confederates would today not care to have such wild-oats sowing made public, though there was nothing but boiled oats, milk and sugar in the mess. Mr. Justice Holmes,[5] however, will not, I believe, take it ill that we are proud to remember his membership; nor will Joseph Warner, Esq.[6] Nicholas St. John Green[7] was one of the most interested fellows, a skillful lawyer and a learned one, a disciple of Jeremy Bentham. His extraordinary power of disrobing warm and breathing truth of the draperies of long worn formulas, was what attracted attention to him everywhere. In particular, he often urged the importance of applying Bain's definition of belief,[8] as "that upon which a man is prepared to act." From this definition, pragmatism is scarce more than a corollary; so that I am disposed to think of him as the grandfather of pragmatism.[9] Chauncey Wright, something of a philosophical celebrity in those days, was never absent from our meetings.[10] I was about to call him our corypheus; but he will better be described as our boxing-master whom we—I particularly—used to face to be severely pummeled. He had abandoned a former attachment to Hamiltonianism to take up with the doctrine of Mill, to which and its cognate agnosticism[11] he was trying to weld the really incongruous ideas of Darwin.[12] John Fiske[13] and, more rarely, Francis Ellingwood Abbot,[14] were sometimes present, lending their countenances to the spirit of our endeavours, while holding aloof from any assent to their success. Wright, James, and I were men of science, rather scrutinizing the doctrines of the metaphysicians on their scientific side than regarding them as very momentous spiritually. The type of our thought was decidedly British. I, alone of our number, had come upon the threshing-floor of philosophy through the doorway of Kant, and even my ideas were acquiring the English accent.[15]

Another reference by Peirce to the "Metaphysical Club" occurs in a letter of 1905 to his former student at Johns Hopkins, Mrs. Ladd-Franklin:

> It must have been 1857 when I first made the acquaintance of Chauncey Wright, a mind about on the level of J. S. Mill. He was a thorough mathematician of the species that flourished at that time, when dynamics was regarded (in America) as the top of mathematics. He had a most penetrating intellect. There were a lot of superior men in Cambridge at that time. I doubt if they could be matched in any other society as small that existed at that time anywhere in the world. Wright, whose acquaintance I made at the house of Mrs. Lowell, was at that time a thorough Hamiltonian; but soon after[16] he turned and became a great admirer of Mill. He and I used to have long and very lively and *close* disputations lasting two or three hours daily for many years. In the sixties I started a little club called the Metaphysical Club. It seldom if ever had more than half a dozen present. Wright was the strongest member and probably I was next. Nicholas St. John Green was a marvelously strong intelligence. Then there were Frank Abbot, William James, and others. It was there that the name and doctrine of pragmatism saw the light.[17]

A letter of Peirce's "To the Editor of the Sun" (n.d.), under the caption "Pragmatism Made Easy," [18] throws a significant light on the reason why Peirce associated the genesis of pragmatism with a *group* of thinkers rather than with the name of any one individual. From historical examples of the discovery of a new idea in the sciences—for example, the conservation of energy and the periodic table—Peirce concludes in this letter that "it is almost self-evident that simply to assign the idea to an individual can give little account of what the process was that actually took place." The social nature of thought is an essential part of Peirce's evolutionist philosophy and enters into his definition of truth. Unlike Dewey, however, Peirce scarcely deals with any specific social problem. He is glad to have known "something of the inwardness of the early growth of several of the great ideas of the Nineteenth Century. By far the most interesting of these was the idea of pragmatism." [19] It was only after he gave ten years to the precise, systematic, and scientific study of philosophy that Peirce offered any contributions of his own, in the *Journal of Speculative Philosophy* (1868).[20] Three years later he produced a more elaborate set of metaphysical categories which he found illustrated in the logic of the sciences, based on what he had learned from his own scientific work and contacts with "what certain students at once of science and of philosophy were turning in their minds," as his letter to the Editor of the *Sun* continues. Then, without naming the Metaphysical Club, Peirce goes on to refer to the members of that group as follows:

> After my return [from Europe, where he had interviewed the scientific students of philosophy], a knot of us, Chauncey Wright, Nicholas St. John Green, William James, and others, including occasionally Francis Ellingwood Abbot and John Fiske, used frequently to meet to discuss fundamental questions. Green was especially impressed with the doctrines of Bain, and impressed the rest of us with them; and finally the writer of this paper brought forward what *we* called the principle of pragmatism.

I have italicized the *we* because it is certain that nobody else in the group but James and Peirce used the name "pragmatism" in print, and that was not until 1898.

On the back of a letter dated March 24, 1907, there is an unpublished fragment of an interesting description, in Peirce's handwriting, of what is evidently the Metaphysical Club:

> —a name chosen to alienate all whom it would alienate. Its constitution was equally effective, for it consisted in a single clause forbidding any action by the Club as a collective body, thus preventing it from wasting the only intrinsically precious element in the world, as so many other societies waste it, in the idle frivolity they call "business," while moreover since without action there could be no officers and in particular no secretary and so no acknowledged record of debate, to gentlemen desirous of distinguishing themselves or of taking patents as it were upon such ingenious combinations of ideas as they might contrive, an adequate motive was presented to hold their peace and abandon the arena of debate to those who only sought to draw as near to the truth as they could. It was quite the most successfully organized body of students I ever had the benefit of joining,—a model worthy of imitation. One of us was Chauncey Wright.[21]

Justice Holmes in a letter (July 21, 1920) to Professor Cohen (who was then writing a series of articles on American philosophy, which included the pragmatists Peirce, James, and

Dewey) wrote: "As to pragmatism I must quote something I said *in 1891 before I ever heard of it*" [22] (my italics). Now, according to Peirce's first account above, Holmes was a member of the Metaphysical Club when it met in the early seventies. Peirce also claims he used the term "pragmatism" frequently in the Club's discussions. It follows, therefore, that either Holmes was absent or inattentive when Peirce used the term, or that his or Peirce's memory is at fault. In any case, Peirce cannot claim to have impressed the term on the minds of any members of the Club with the single exception of William James, and none of the members, including James, and none of their friends ever recorded the name of the Club which looms so large in Peirce's accounts of the genesis of pragmatism. Justice Holmes did inform Professor Cohen that he suspected Chauncey Wright was the source of Peirce's tychism. [23]

The only other reference to the Club that I have been able to find in Peirce's writings is in an article published as late as 1908. It provides further evidence that Peirce wished to distinguish his *logical* principle of pragmaticism from James's doctrine.

> In 1871, in a Metaphysical Club in Cambridge, Massachusetts, I used to preach this principle [of pragmaticism] as a sort of logical gospel, representing the unformulated method followed by Berkeley, and in conversation about it I called it "Pragmatism" . . . Of course, the doctrine attracted no particular attention, for . . . very few people care for logic. But in 1897 Professor James remodelled the matter, and transmogrified it into a doctrine of philosophy, some parts of which I highly approved, while other and more prominent parts I regarded, and still regard, as opposed to sound logic. [24]

Peirce's analysis of meaning and of the method of making ideas clear by considering their *logical* consequences and *general* or habitual effects on our conceptions, was interpreted by James in *psychological* and *individualistic* terms. Hence, Peirce distinguished James's pragmatism from his own, which he baptized "pragmaticism," [25] and then went on to observe that James "no doubt derived his ideas on the subject from me." [26] This latter claim has led Professor Perry properly to raise the "nice question whether it is possible to 'derive' from a philosopher ideas which he has never had; or whether one may not reasonably doubt the paternity of a bantling which, as it grows older, becomes increasingly dissimilar to its father. Perhaps it would be correct, and just to all parties, to say that the modern movement known as pragmatism is largely the result of James's misunderstanding of Peirce." [27] No wonder that Peirce wrote to James in 1900: "Who originated the term 'pragmatism,' I or you? Where did it first appear in print? What do you understand by it?" Professor Perry thus significantly notes: "Though the origin of pragmatism be obscure, it is clear that the idea that pragmatism originated with Peirce was originated by James." [28]

The chief historical source of Peirce's definition of pragmatic belief is Kant's *Critique of Pure Reason*, a work well known and discussed at great length by Wright and Peirce. The following passage from Kant's work, if we disregard the other transcendental doctrines of Kant, illustrates and defines what any of our pragmatists would have accepted as a characterization of all beliefs:

> The physician must do something in the case of a patient who is in danger, even if he is not sure of the disease. He looks out for symptoms and judges, according to his best knowledge, that it is a case of phthisis. His belief is even in his own judgment only a contingent one; someone else might perhaps judge better. I call such contingent belief which still forms the basis of the actual use of means for the attainment of certain ends, *pragmatic belief*.
>
> The usual touch-stone or test of whether something is just talk or at least subjective conviction, that is, firm belief, is the *bet* . . . A bet makes one stop short . . . If in our thoughts we imagine the happiness of our whole life at stake, our triumphant judgment disappears, we tremble lest our belief has gone too far. Thus pragmatic belief has degrees of strength varying in proportion to the magnitude of the diverse interests involved.[29]

Holmes's bet-abilitarianism, we shall see, is an instance of the sort of pragmatic belief defined by Kant, but the great difference between the American pragmatists and Kant is their denial that over and above contingent pragmatic belief are the purely rational, necessary, and absolute ideas of Kant's transcendental philosophy.

Peirce rejected the absolutistic sense of the term as it occurs in Kant's a priori ethics of "praktischen Vernunft" and its categorical imperative. He preferred the Kantian term "pragmatisch" which stood for the humbler means-ends relation expressed in hypothetical imperatives, for example, counsels of prudence which sanction laws by their consequences for the general welfare, or the teachings of history. "If you want to enjoy security in society, enforce the laws," is a "merely" pragmatic rule for Kant, to be sharply distinguished from the purely ethical commands of a disembodied and rational free will. Now, since all reasoning, scientific or ethical, was for Peirce hypothetical and operational, that is, guided by conceivable empirical consequences, he preferred to name his method "pragmatism" rather than "practicalism."

For one who had learned philosophy out of Kant, as the writer, along with nineteen out of every twenty experimentalists who have turned to philosophy, had done, and who still thought in Kantian terms most readily, *praktisch* and *pragmatisch* were as far apart as the two poles, the former belonging in a region of thought where no mind of the experimentalist type can ever make sure of solid ground under his feet, the latter expressing relation to some definite human purpose. Now quite the most striking feature of the new theory was its recognition of an inseparable connection between rational cognition and rational purpose; and that consideration it was which determined the preference for the name *pragmatism*.[30]

So far James and Peirce were agreed. They parted company on the doctrine of the will to believe, central to James's pragmatism but criticized sharply by Peirce, Wright, and Holmes. We shall see later how Peirce attempted to convert the Darwinian ideas of chance variation and natural selection into the idea of an evolution of the mind by means of a logical competition among thoughts, which eliminates ideas not fit to stand for the truth fated to be discovered by those who investigate. The methods of authority, tradition, and tenacity compete with that of science to fix belief in every walk of life. The exact sciences were evolving to a stage where the rules for the efficient discovery of laws were beginning to take shape. In psychology, history, law, as well as in metaphysics, a growing consciousness of method was apparent to Peirce, and the living manifestations of the "growth of concrete reasonableness" were found by him in the thinking of the distinguished associates upon whom he conferred membership in his Metaphysical Club. If this Club were not primarily a symbol in Peirce's metaphysical imagination of the "Search for a

Method"—as he called one of his many unfinished treatises—one would expect to find at least one reference to the Club in the writings or biographies of its members. However, there is no mention of this Club in James's writings, filled as they are with generous acknowledgments.

Nor do we fare better if we consult the letters and biographical testimonials of the senior and leading member of the Club, Chauncey Wright. For example, Joseph B. Warner, who attended Wright's class in Psychology at Harvard in 1870, and assisted Oliver Wendell Holmes, Jr. in his edition of Kent's *Commentaries on American Law*, does not mention the Club in his contribution to Thayer's *Letters of Chauncey Wright, With Some Accounts of His Life*. More than half a century after Wright's death (1875), Justice Holmes recalls Chauncey Wright's probabilism[31]—an essential element of Peirce's pragmaticism—but never mentions the Metaphysical Club. The letters of John Fiske, gossipy as they are about old Cambridge and his personal relations to the leading lights of the Harvard Yard, yield the same negative result.[32] The intimate letters of Chauncey Wright's Northampton and Cambridge friends, E. L. Gurney and Charles E. Norton, and the lengthy Wright-Abbot correspondence fail to refer to the Club. Perhaps some further search may secure more data than my unrewarded efforts to supplement Peirce's history of a Metaphysical Club whose existence is so definitely associated in his mind with the birthplace and conscious adoption of a new philosophical doctrine. Either Peirce invented the name of the Club, "half-ironically, half defiantly," or the group he named was a more casual and informal one than anything one would call a club. We do not know who, besides James, heard or read Peirce's paper, published about six years later in two articles, "The Fixation of Belief" and "How to Make Our Ideas Clear," in the *Popular Science Monthly* (1877-78), but the lonely Peirce tells us more than thirty years later that it was in that group that "the name and doctrine of pragmatism saw the light." The undeniable fact is that no term like "pragmatism" occurs in either article or in the French version that appeared in *La Revue philosophique* in 1879.

But, after all, what's in a name? The most significant fact for the historian of thought is that Peirce brought together in his account of the genesis of pragmatism a historically important group of persons who really lived in the same place and time, moved in the same intellectual atmosphere, and influenced each other in ways that shaped the growth of certain pervasive ideas current in our thinking today. Peirce was more conscious than any of his contemporaries of the historical significance of the discussions by his group of the major ideas of the time, particularly the idea of evolution, or, more accurately, *evolutionism*—the generic name for the flock of generalizations that invaded every province of thought with the gradual acceptance of the Darwinian theory of evolution, much debated in the two or three decades after 1859. Many of these provinces were represented by the several members of Peirce's "Club." From the distinctive bent each gave to the idea of evolution in his chosen field of work there emerged a dazzling variety of pragmatisms.

Only two of the members of the Club, Peirce and James, used the term "pragmatism" as the name for a way of settling metaphysical puzzles—to few metaphysicians' satisfaction—but all of them exhibited in their more concrete fields of work a more fruitful use of a pragmatic *method* capable of diverse applications to the clarification of the basic concepts of the natural and social sciences. That the meaning of a theory *evolves* with its experimental applications, that all claims to truth have to be publicly verifiable and withstand the competition of prevailing ideas, and that the function of ideas is to adjust man to a precarious and changing world—these are essential aspects of the evolutionary meaning of the method of pragmatism. We shall watch that meaning grow with its applications by the members of the Metaphysical Club in their several fields. Different aspects of the complex relation between evolutionism and the growth of pragmatism will appear in the ensuing chapters on the specific contributions and philosophical attitudes of each of the members. They form two groups, one trained in the natural sciences (Wright, Peirce, James), the other in historical and legal studies (Fiske, Green, Holmes, Warner); but all were interested students of philosophy and animatedly discussed the broader implications of evolution. Out of the cross-fire of their opinions about the prevailing claims of metaphysical and religious interpretations of evolution, one type of question emerged more clearly and urgently than all the rest: just *how* was one to proceed in thinking through the tangle of scientific, ethical, religious, and metaphysical ideas about evolution? It became clear to the founders of pragmatism in the midst of the momentous debate of their age that a rule of *method* was required for fixing one's beliefs, for making one's ideas clear, for testing intuitions, traditional and a priori modes of thinking, for verifying the fruitfulness and concrete bearings of all general ideas about thought and conduct in the light of their consequences for the sciences and for the welfare of man.

The two main schools of philosophy, British empiricism and German idealism, had to be overhauled and reappraised in the pragmatic search for a method. Locke and his followers had more than suggested "a plain historical method"—that is to say, a psychological method—of testing ideas by comparison with particular experienced effects. Kant had formally pointed out the regulative character of the ideas of pure reason, and Hegel had elaborated upon the transformative power of reason in the history of civilization. But it was the unevolutionary character of old world forms of static sensational empiricism and rigid a priori rationalism which provoked the pragmatists at Harvard to sense the intellectual need of a more flexible and dynamic view of the nature of experience and of the function of reason. They rejected the British theory of the mechanical association of ideas as well as the German fantasy of the "ballet of bloodless categories." Reason was not merely Hume's slave of the passions nor Hegel's absolute lord of creation; it was an instrument which had evolved from animal cunning to become the sole reliable means of attaining the free use of one's natural powers. In social matters it was the only workable means of achieving a coöperative mode of living with others who had competing desires. Studies of the history of early societies by Tylor, Maine, and others, had plainly shown that predatory roots and barbaric customs were the evolutionary bases of the law and other social institutions.

The more optimistic historian, John Fiske, was only an interested friend of the pragmatic members of the Club, but his attempted idealistic, evolutionary synthesis of the physical and social sciences along Spencerian lines ran counter to the realistic pluralism of the scientific and legal members. Fiske's sociological philosophy of history in his *Outlines of Cosmic Philosophy* and his controversy with William James on the evolutionary role of great men in history, will show* how diverse evolutionisms affected the social sciences in the 1870's just before the emergence of a more pragmatic legal and social philosophy. "In 1869, the New York *World* reported in full Fiske's Harvard lectures [on Evolution]. There has never been any other case in American history of such a popular interest in a philosophical controversy. . . . He traced the growth of American institutions as Darwin traced the growth of species." [33]

Historians and social scientists today may look suspiciously at the clumping together of so many questions and ideas belonging to specialized provinces of research. But I think we must admire the philosophical scope and grandeur of vision of the founding fathers of pragmatism. They had a truly philosophical interest in such important scientific discoveries of their age as the general laws of energy-transformation ("correlation of forces"), the evolutionary development of the earth's crust and of living species, the primitive forms of society, ancient laws and institutions, the evolution of consciousness, and the historical synthesis of all these laws of nature and society. The historical or genetic, temporal aspect of all observable phenomena and ideas dominated the foreground of the whole evolutionary and early pragmatic approach to nature and man. It would be doing historical violence and philosophical injustice to the ideas of our evolutionary minded pragmatists to dismiss their historical approach to scientific and philosophical problems as a simple case of the genetic fallacy. There is ample evidence that they were aware of the practical and logical need for distinguishing between the historical and the analytical approaches to problems in their respective fields. They were also, without exception, quite critical of pseudo-scientific evolutionary syntheses like Herbert Spencer's, as well as of antiscientific dialectical systems like Hegel's.

There is no one coherent theory of evolution or of reality to be found in the writings of any one, or of all, of the pragmatic thinkers discussed in this book. On the contrary, we are indebted to them for their acute sense of the overweening pretentiousness of all such systems, and of their failure to account

* See Chapter VI below.

for the temporality, complex diversity, relativity, and individuality of natural and social phenomena. Running through all our early pragmatic thinkers is a strong sense of liberation from the facile monisms of both materialistic and spiritualistic systems of evolutionary metaphysics and theology. This pragmatic proclamation of intellectual freedom in the midst of the evolutionary controversy sprang from two powerful sources: the rapid growth of the physical, biological, and human sciences, and the democratic Anglo-American tradition upholding the sacredness of the rights of individuals. If we separate the scientific from the political aspects of liberalism we get a narrow positivism and isolated individualism impotent to deal with the institutional aspects of science and society. If we fuse the two uncritically we violate the neutrality of science (the key to Chauncey Wright's philosophy) and limit the freedom of inquiry. This dilemma runs as a deep undercurrent in the liberalism of the founders of pragmatism. It was manifest in their reactions to the excesses of Spencer's sociological evolutionism. They fathomed the dilemma's depth and brought to the surface intellectually and humanly valuable insights, summarized in our concluding chapter as their philosophical legacy to the twentieth century. Their specific investigations of the foundations, methods, and goals of the natural and social sciences furnish concrete evidence of the fruitfulness of their pluralistic and humanistic approach to the complex problems of civilization. Though it was a common failing in the latter half of the nineteenth century for evolutionary philosophers like Fiske to confound historical and logical problems in an over-optimistic faith in the inevitability of progress, none of our pragmatic thinkers failed to criticize the prevailing belief in automatic progress guaranteed by infallible dogmas or inflexible traditions. An experimental, tentative, and flexible approach to the problems of the natural and social worlds is an essential characteristic of American liberalism. How a variety of pragmatisms evolved in the diverse fields of study of the liberal members of the Metaphysical Club out of the ferment of the evolutionary controversy of the last century, is the central theme of this book.

The conscious adoption of pragmatism as "a method in aid of philosophic inquiry" and in the discussion of the dark questions of metaphysics, was associated by Peirce with the ideas dis-

cussed in his Metaphysical Club. In order to trace the history of
the idea of pragmatism in the United States, and of some of the
thirteen or more meanings it had acquired by 1908,[34] we shall
have to consider the effects on the members of the Club of their
discussions eventuating in diverse developments of the most stirring theme of the third quarter of the nineteenth century, evolution. The subsequent history of evolutionism is intimately linked
with the genesis of pragmatism as a newly and diversely formulated doctrine. The growth of that doctrine out of the scientific
and philosophic background in the United States will be sought
in the divergent interpretations of evolutionism by the members
of Peirce's Metaphysical Club. Fortunately for the historian of
ideas, they were brought together in the mind of one so profoundly steeped in the history of philosophy as Charles S. Peirce,
who after forty years of dissecting thought, may well have come
to realize the wisdom of Chauncey Wright's words:

> The most profitable discussion is, after all, a study of other minds,—
> seeing how others see, rather than the dissection of mere propositions. The
> re-statement of fundamental doctrines in new connections affords a parallax
> of their philosophical stand-points (unless these be buried in the infinite
> depths), which adds much to our knowledge of one another's thought.[35]

Wright's statement has been the guiding principle of our
own study of the minds of the founders of pragmatism. It may
well serve as a pragmatic formulation of the essential aim and
method of all intellectual history when liberally conceived as
part of the larger pursuit of philosophical and humane understanding. In that spirit we turn to the scientific and philosophical
luminaries of the Metaphysical Club for the purpose of taking the
parallax of their pragmatic standpoints whenever visible through
the infinite depths.

CHAPTER THREE

Chauncey Wright, Defender of Darwin and Precursor of Pragmatism

1. THE MIND OF WRIGHT, "CORYPHAEUS" OF THE METAPHYSICAL CLUB

Chauncey Wright was an able mathematician, and though by trade an astronomical computer, his intellectual interests were varied. He emphasized the importance of understanding the methods of exact sciences, and of geology, biology, psychology, sociology, ethics, and metaphysics. He was one of the first after Asa Gray to defend Darwin's theory of natural selection, and to publish applications of that theory to the biological problems of explaining on physical principles the building instincts of bees, and the distribution of leaves around a stem or plant-axis (phyllotaxis). These contributions of Wright, limited in scope though they were, were appreciated by his teacher Asa Gray.[1] When Darwin, apparently impressed by Wright's analytic powers, asked him to make clear when a thing may "be properly said to be effected by the will of man," [2] Wright set to work on what turned out to be his major contribution to scientific psychology—or, as he called it, psycho-zoology—"The Evolution of Self-Consciousness." [3] Evolution, argued Wright, had to be restricted to biology and psychology, for the geologic record was too incomplete—and the astronomical record even more so—to justify evolutionary theories of the physical history of the world. Cautious methodological pluralist that he was, Wright set his face against those who speculated metaphysically about evolution in history and social ethics.

In short, Wright was not merely a defender of Darwin. Despite his modesty, he boldly set himself against nearly all the prevailing philosophies of evolution of his time. Under the banner of the neutrality of science he condemned as futile the efforts of idealists, materialists, and Spencerians to appropriate the latest achievement of science, Darwin's theory of evolution, for their own ends. He argued that no scientific theory could logically support outworn ethics and social sanctions with which it had nothing to do. When Wright argued thus for the ethical and metaphysical neutrality of science, he may be said to have continued Kant's insistence upon separating scientific or theoretical questions from ethical or "practical" ones. Wright, starting from Darwin's work, came to conclusions similar to those of Kant's critical philosophy which dealt with the implications of Newton's physics. Without anything like the architectonic system of Kant, Wright produced the most level-headed discussions of his time concerning the philosophical meaning of Darwin's work. Though his philosophical defense of Darwin was admired by his friends at Harvard, he met with opposition from some professors of philosophy and natural theology. One of them was Francis Bowen, Alford Professor of Moral Philosophy at Harvard (from 1853 to 1889), an adversary of Dr. Oliver Wendell Holmes on the relation of science to theology; he wrote four letters to Wright criticizing his first paper in support of Darwin, "Remarks on the Architecture of Bees." [4] Bowen insisted on a transcendental distinction between physical and metaphysical causality, urging that empirical sciences can deal only with inert things and furnish only probability as a measure of our ignorance, whereas God alone can be the real, productive, and necessary cause of living forms. Bowen's discussion of probability and causality in these letters, and in his many other idealistic works, is so confused by extraneous theological and teleological ideas that Wright's quiet scientific patience must have been sorely taxed. But Bowen's sort of theological argument was then widespread in respectable scientific and academic circles, as I have indicated in the first chapter. Theism enjoyed the protection and scientific prestige of the foremost biologist in America, Louis Agassiz, Professor of Zoology at Harvard, and teacher of Wright, Peirce and James. Wright was thus in the midst of the major intellectual movement of his day, the impact of Darwinism on the logic of science, religion, ethics, and sociology; and so his thought reveals a great deal about the intellectual history of the times and of the other members of the Metaphysical Club. His conversion to Darwinism immediately after reading the *Origin of Species* when it first came out, his welding it with Mill's empiricism, his critique of Spencer's and Fiske's evolutionism, his correspondence with Francis Ellingwood Abbot,° his philosophical discussions with James on "psycho-zoology" and the "will to believe," with Peirce on Kant, Mill, and Darwin, with Green and Holmes on dispensing with metaphysical necessity in the law, "his educative influence on all those who enjoyed his intimacy" in and around Harvard Yard, make him, in short, as Peirce reports, the central figure of the Metaphysical Club, "where the name and doctrine of pragmatism saw the light." Wright is unmistakably our key figure, even though he never wrote a book and never used the term "pragmatism."

When Darwin's epoch-making work appeared in the United States, Chauncey Wright was teaching natural philosophy in Professor Louis Agassiz's school for girls in Cambridge. On February 12, 1860, he confided to his friend Mrs. Lesley, to whose mother he was indebted for having successfully urged his father to send him to Harvard, that he had just finished reading

that new book on "The Origin of Species,"—Darwin's,—to which I have become a convert, so far as I can judge in the matter. Agassiz comes out against its conclusions, of course, since they are directly opposed to his favorite doctrines on the subject; and, if true, they render his essay on Classification a useless and mistaken speculation. I believe that this development theory is a true account of nature, and no more atheistical than that approved theory of creation, which covers ignorance with a word pretending knowledge and feigning reverence. To admit a miracle when one isn't necessary seems to be one of those works of supererogation which have survived the Protestant Reformation, and to count like the penance of old for merit in the humble philosopher. To admit twenty or more (the more, the better), as some geologists do, is quite enough to make them pious and safe. I would go farther, and admit an infinite number of miracles, constituting continuous creation and the order of nature.[5]

Charles Peirce, on returning from his surveying in Louisiana about June 1860, had noted that Wright, after abandoning Hamilton's Kantianism for Mill's nominalism, "was now all enthusiasm for Darwin, whose doctrines appeared to him as a sort of supplement to those of Mill." [6] Wright had none of Peirce's metaphysical

° See pp. 42–48.

qualms about the nominalism of the British empiricists, and defended Darwin's work as evidence of the fruitfulness of the empirical tradition upheld by a long line of British philosophers: Occam, Francis Bacon, Boyle, Newton, Locke, Berkeley, Hume, Bentham, and Mill. We recall that Peirce confessed himself to be the only member of the Metaphysical Club whose philosophic approach was not at first oriented by British empiricistic nominalism, since he had early taken up with Kant, Schelling, and Scotistic realism. On the other hand, Wright was proud to be descended intellectually and by family ancestry from the English, as he again confided to Mrs. Lesley when Darwin's *Descent of Man* appeared in 1871 with acknowledgments of Wright's work:

> I grow more and more conscious every year that my most cherished thoughts and interests are of English origin. My blood, though English too, is nothing to them but their accidental road. All American interests and charm are in the future, or in a short development prophetic of this future. They are not new powers or principles, but better opportunities for the old to work themselves out. They are not the struggle for the victory, but the realization of its fruits. For five hundred years, from the time when old William Occam asserted common sense and experience against the devoted and enthusiastic subtilties of continental and Celtic schoolmen, England has taken the lead in every great revolution and practice, even down to Darwinism. Other nations have done much in carrying out and even in discovering in detail the principles of practice and science; but wherever a great victory had to be won for progress, and principles had to be established not only in experience but against authority, English genius has done it. If this be attributed to English freedom, it comes to the same thing; for English freedom was the product of English genius or common sense, aided, no doubt, by an insular position.[7]

Thus, to Wright's genetic way of thinking it was not a historical accident "that Mr. Mill is the modern champion at once of nominalism in logic and of individualism in sociology." [8]

The traditional English respect for the facts of particular experiences, the "experimental" philosophy of Locke, Berkeley, Hume, and Mill, was, according to Wright, triumphant in Darwin's detailed methods of close and patient observation, despite the hue and cry raised by theological-minded opponents of Darwin. Nor, in Wright's mind, were the diverse interests of religion and science aided by metaphysical or theological defenders of evolution. The Scottish "common-sense" realists, represented at

Princeton by James McCosh,[9] failed, according to Wright, to do justice either to practical morality or to theoretical science: by appealing to innate intuitions they left no room for the historical evolution of moral conscience or self-consciousness. Mill's attack upon Hamilton in 1865 and Mansel's weak defense led Wright, an admirer of Mill, to an even more positivistic defense of science. Even in Germany, the idealistic neo-Kantian philosophies of Fichte, Schelling, Hegel, and Schleiermacher were being attacked by materialistic critics of Hegel (Strauss, Feuerbach, Marx) and by other scientific materialists, Moleschott, Büchner, and Oken. But Wright was repelled by the dogmatic metaphysical pretensions of both the idealists and the materialists. He maintained steadily that neither side could logically affect the experimental and inductive method of science.

Finally, Wright strongly condemned Herbert Spencer's *Synthetic Philosophy* (1860) and Fiske's version of it for reading into evolution a dubious system of inevitable moral and theological interpretations in the fashion of what Wright toward the very end of his life called "German Darwinism." [10]

Wright was much more cautious than the other members of the Metaphysical Club in interpreting the philosophical significance of Darwin's and Spencer's theories of evolution for ethics and metaphysics. We shall say only a word here about the divergent evolutionary ideas of the other members, since they will be treated extensively in the chapters to follow. Peirce maintained that philosophy had to be based on a thorough-going evolutionism or none at all, though he refused to approve of Darwin's theories to the exclusion of Lamarckian and catastrophic theories of evolution, especially in accounting for the history of thought and civilization. In that history Peirce's metaphysical evolutionism saw "the growth of concrete reasonableness" in the universe at large. John Fiske drew large audiences with his lectures on cosmic philosophy, in which he offered his theory of man's prolonged infancy as an "original" contribution to the theory of evolution. He saw in human history *The Destiny of Man, Viewed in the Light of His Origins,* and later applied this vision to his researches in American history. Abbot early accepted Darwin but interpreted his theory of evolution in metaphysical and theistic terms on the basis of his "scientific realism." Holmes and the other lawyer members of Peirce's Club laid the basis of modern legal realism by regarding the law as the product of evolving sociological ·factors. There seems to be no doubt that what James condemned as amoralism in Holmes and a kindred nihilism[11] in Wright clashed with the moralistic strain in William James, especially at a time when the latter was just emerging from the depths of a mental depression and had turned to the fideism of Lequier and Renouvier.[12] James criticized Wright's nihilism in an article perhaps read and discussed at a meeting of the Metaphysical Club, and Wright defended himself in critical notes written on this manuscript.[13] Underlying the candidly and often sharply expressed differences on the relation of science to religion, among the members of the club there was a common concern to preserve what was valuable in each.

A conversation with Wright on "Living according to Nature," taken down by one of his friends in January 1875, shows a much sharper distinction between cosmic and moral nature than one finds in James, Peirce, Abbot, or Fiske:

> It is permissible to use the word Nature as the name of the harmony of things, but it is not permissible to confound the harmony in the whole, the laws of nature and the invisible orders both without and within us,— to confound the law of causation, whose formula is, "If thus, then so," with the harmony we seek as moral beings, which without our seeking would not, and does not exist . . . The laws of this [moral] harmony are of a wholly different order, *different in meaning*, out of the other's sphere, neither contradictory to nor in conformity with those of the scientific cosmos: though involving them as the laws of living structures involve those of matter generally, or as the laws of mechanical structure involve those of its materials and surrounding conditions. Mechanical structures, living structures, artistic and moral structures, are all fittings to ends; and these though not absolute accidents (since nothing in the cosmos is absolutely accidental), yet relatively to any discoverable principles of the cosmos, are accidents. Now, the conditions which determine these several forms of fitness are in the cosmos, but the ends are not,—except so far as human imaginations preconceive them, or as the actualities of constructions in living forms in art and in moral character are their embodiment . . . The way to follow Nature is to observe the means which, in accordance with the cosmic laws or conditions of Nature, and of human nature, are found to be conducive to self-sanctioned ends in the higher social or moral life of man, or in his reflective social nature . . .
>
> While, therefore, it is not permissible, in respect to the harmony of ideal ends with the outward activities of life, to confound it with those

laws of universal nature that are not to be obeyed, since they cannot be violated; yet the theoretical fault of this confusion is in some sort compensated by the practical value and force it has had with many minds of the poetic type . . . Reverence, or at least the poetical form of it, demands that power and goodness or moral harmony should exist *in actu*, in a being in real nature, as well as *in posse* or ideally. Historically, this tendency has been of the greatest service to moral advancement.[14]

We find Wright's distinction between the aims of science and those of metaphysics reflected at times in the writings of James and Peirce, as we shall have occasion to notice in the next two chapters. In his *Principles of Psychology,* James found Wright's psycho-zoological "Evolution of Self-Consciousness" correct in excluding metaphysics from scientific inquiry. Each science assumes as its data elements whose ultimate metaphysical grounding is not the concern of scientists. Their task is confined to investigating the operations and relations of these elements in diverse observational contexts. This positivistic strain is encountered in Peirce also:

It does not belong to the function of a scientific man to ascertain the metaphysical essence of laws of nature. On the contrary, that task calls for talents widely different from those which he requires. Still, the metaphysician's account of law ought to be in harmony with the practice of the scientific man in discovering the laws; and in the mind of the typical scientific man, untroubled by dabbling with metaphysical theories, there will grow up a notion of law rooted in his own practice.[15]

Chauncey Wright once told some friends at a party that metaphysics was a favorite amusement of his, solving mathematical problems his favorite occupation. We know that his nihilistic theory of metaphysics disturbed those of his philosophical friends who expected from metaphysics some answer to the questions: Why do we exist? What is the value and purpose of life? Such questions, Wright proposed, should be analyzed, freed from ambiguities concealed in vague terms like "life," and considered empirically:

I do not feel so confident about your problem, "Why do *we* exist?" . . . All the ends of life are, I am persuaded, within the sphere of life, and are in the last analysis, or highest generalization, to be found in the preservation, continuance, and increase of life itself, in all its quantities of rank,

intensity, and number which exists—"for what," do you ask? Why, for
nothing, to be sure! Quite gratuitously. Does any one seriously expect to be
answered in any other terms than those in which the question could be
rationally framed? Are any ends suggested out of the sphere of life itself?

The social value of questions is, indeed, a matter we might overlook
in a too serious purpose to find their answers. The social value of the
weather is nothing to theirs; and insoluble questions have a permanent
value of this sort. Religions are founded on them.

Still in the interest of sober inquisitiveness, it might be worth while
to root out some of these questions for the sake of others more genuine.
Let the questions of the uses of life, then, be put in this shape: To what
ascertainable form or phase of life is this or that other form of life valuable or serviceable? . . . I am more than half persuaded that most, if not
all, of the puzzles of metaphysics, may be reduced to unconscious puns,
or unseen ambiguities in terms. Now, "life" means in common discourse
two things very different, but easily confounded. We sometimes mean by
"life" what is comprised in the plans, purposes, inquisitions, and aspirations that make ambition and the zests of curiosity and anticipation so large
a part of the conscious life of youth. Well, if for any cause (an indigestion,
for example), the strength and zest that went in search of these goods of life
happen to fail, we say we have tried life, and exhausted its resources! It has
no more value for us. We are ready to die! If we meditate suicide, it is
not our duty to others, nor the rights others may have in our lives, that
should restrain us. We are more irrational than to merely forget the claims
of conscience. We suffer from a mental indigestion. We have not solved
the ambiguities of words. The *life* we would attack is not the culprit. The
kind of dying which a wise moralist would enjoin us is the death of the
unsatisfied anticipations, curiosities, ambitions, which, fixed as habits, still
linger, and distress the soul, since the strength and zest have failed which
could give then further fruition. But these forms of life die without sacrificing one's usefulness or duties to others, or without cutting off a host of
resources which come in old age to make "life" quite tolerable . . . But
the true philosophical way is to look on life as it is, as somewhat broader
than the fading pictures, plans, and purposes which you have mistaken for
it, and as consisting in more than that set of inveterate habits which you
call yourself. The death you should desire is the death of those desires,
which, like all unsupported or no longer satisfied impulses of habit or instincts have become pains. Work in other channels, and thus immolate
yourself, and you will not find an end of life desirable. Life in this wider
sense is neither good nor evil; but the theatre of possible goods and evils.[16]

Such dispassionate objectivity in Wright's philosophical attitude led William James to characterize him as follows:

Never in a human head was contemplation more separated from desire. For to Wright's mode of looking at the universe such ideas as pessimism

or optimism were alike simply irrelevant. Whereas most men's interest in a thought is proportional to its possible relation to human destiny, with him it was almost the reverse. When the mere actuality of phenomena will suffice to describe them, he held it pure excess and superstition to speak of a metaphysical whence or whither, of a substance, a meaning, or an end . . .

He particularly condemned the idea of substance as a metaphysical idol. When it was objected to him there must be some principle of oneness in the diversity of phenomena—some *glue* to hold them together . . . he would reply that there is no need of a glue to join things unless we apprehend some reason why they should fall asunder. Phenomena *are* grouped—more we cannot say of them.[17]

James offered his "will-to-believe" doctrine as an antidote for what he dubbed the "anti-religious" teaching of this neutral scientific skepticism and "nihilism" of Wright's, and had put forward his views in the paper "Against Nihilism."[18] Wright's reply to James defines nicely and profoundly the philosophical position which made him such "an educative influence upon the minds of all of us who enjoyed his intimacy," as Peirce acknowledged. "After all," Wright argued, "nihilism is rather a discipline than a positive doctrine; an exorcism of the vague; a criticism of questions which by habit have passed beyond the real practical grounds of causes of question. Common sense is opposed only so far as common sense is not critical."[19]

Both Peirce and James acknowledged their personal debt to Chauncey Wright's stimulating talk and participation in the discussions of the Metaphysical Club. James said that Wright did his "best work in conversation,"[20] and Peirce confesses to the dialectical "pummeling" he received at the hands of Wright, whose fidelity to his neutral and "nihilistic" method of naturalistic empiricism stands in contrast to their less sober, more speculative intellectual ventures. For Peirce was led by his absorption in Duns Scotus and Hegel to some dubious transcendental speculations in which a priori metaphysical categories served as the underpinning of scientific method. It must be said, on the other hand, that Peirce's more solid speculations concerning such topics as the logic of relations and the statistical character of physical laws had a more far-reaching influence on twentieth-century thought than Wright's over-cautious Baconian refusal to indulge in any "anticipations of nature." Critical philosophy ought not to pre-empt the domain of philosophic thought, which may with logical impunity offer speculative views as *hypotheses;* without hypotheses no advances in scientific philosophy are possible.

Chauncey Wright would not have had any influence on American philosophy if he had meant by the neutrality of scientific method to defend a sophisticated form of scornful indifference or blanket skepticism with regard to traditional moral, theological, and metaphysical questions. He was not the sort of scientific specialist who in spare moments permits himself to get drawn into a philosophical discussion about the remote implications of his scientific researches, and then, finding no prospect of a tangible solution, withdraws into his narrower field. There were, of course, minute studies in mathematics, physical theory, botany, zoology, and psychology which Wright undertook while eking out a living as a computer for the Nautical Almanac; but all the biographical evidence* indicates that his predominant interest and favorite topics of discussion were speculative or metaphysical, the latter representing a mixture of logical and psychological problems. What was remarkable about his manner of speculation was its thoroughgoing scientific character, its close attention to the meanings of terms, to the methods of proof employed, and to the many empirical details which differentiate the diverse complexities of nature. His apt expression for this empirical diversity, "cosmic weather," did not imply any inherent ontological indeterminateness in things themselves, but rather a radical pluralism grounded in the *complexity* of nature and in man's many-sided evolved interests, practical, aesthetic, and scientific. Wright assimilated ethics with the practical, religion with the aesthetic, and insisted on regarding scientific and cosmological theories as hypotheses to be purged of any ethical or aesthetic, religious or transcendental bias.

To Wright's empirical and discriminating mind,

religion and morality are both realized in the practical nature of man, in character, in the fashioning of his volitions and desires . . . An irreligious man is, then, first, one who acknowledges no supreme ends or objects; or, secondly, one who though he acknowledges, does not habitually submit his will to such a power . . . An immoral man is, then, one who does not observe habitually and consistently any principles of conduct: either, first, from infirmity of purpose or want of discipline; or, secondly, from pure selfishness

* See Appendix A.

or a disregard of generally desirable ends . . . Religious and moral duties are often indistinguishable. But religious duties, or acts which are believed to be religious duties, come to be distinguished from others by the predominance of two marks: first, their absolute or unconditional character; second, their inutility, or the immediacy of the relation of act to object or end. Besides such immediate and unconditional or religious duties, civilization creates a host of others, and many other motives to their observance . . . The above discriminations are independent of any theory, utilitarian or other, of the source of moral and religious convictions, and are designed to show that religion and morality are definable and distinguishable independently of creeds and codes. This is the same as Mill's position.[21]

These logical distinctions were laid down in a letter to his friend, the practical, positivistic Charles Eliot Norton. Addressing the speculative theologian, Reverend Francis Ellingwood Abbot, Wright adopted a different approach to the same subject, though he still maintained that morals and religion were independent of theoretical or scientific knowledge: "I have always believed that the really *essential* positions of morals and religion could be sustained on the 'lower' ground of common-sense,—on what men generally understand and believe independently of their philosophical theories." [22]

2. WRIGHT'S CRITIQUE OF F. E. ABBOT'S TRANSCENDENTALISM

Wright's way of regarding religion and morals as independent of theological theories, stemmed from the pluralism of the Kantian philosophy of Hamilton which sharply set limits to science and the speculative use of reason in order to make room for the ethics of good will and religious faith. These were the main topics of Wright's correspondence with a theistic and metaphysical critic of Hamilton's Kantism, Francis Ellingwood Abbot, who through his admirer Peirce became linked to the founders of pragmatism as an occasional member of the Metaphysical Club. This correspondence serves to throw into sharp relief Wright's view of the relation of science to ethical and religious values which constituted the central philosophical issue of the Darwinian controversy.

Francis Ellingwood Abbot (1836–1903) was a Harvard classmate of Peirce, who called him "one of the strongest thinkers I ever encountered." [23] On learning of Wright's death in 1875, Peirce proposed to William James a memorial volume: "His memory deserves it for he did a great deal for every one of us. I don't speak of the philosophical *canaille*, but I mean you, Frank Abbot, and myself." [24]

The intellectual relations of Wright to Abbot are found in the series of letters between them beginning in 1864. Abbot was then a respected Unitarian clergyman at Dover, New Hampshire, three years before he began to have difficulties with his more orthodox colleagues which culminated in litigation[25] and abandonment of his profession. He was one of the founders of the "Free Religious Association," [26] a group of positivists who wished to establish their own church. Abbot's *Scientific Theism* was published in 1885 and was followed two years later by a one-year appointment as instructor in philosophy at Harvard, when Josiah Royce was on leave of absence.*

Abbot had written an article on "The Philosophy of Space and Time" for the *North American Review* of July 1864, and had sent a copy of it to Wright for his criticisms, probably because he knew of Wright's one-time espousal of Hamilton's theories. Wright replied (December 20, 1864) that the essay "has pleased me very much by its beautiful philosophical style and admirable clearness, and it seems to me to evince very great metaphysical ability." That Wright was not merely being polite is clear from his repeating this high opinion of Abbot in his subsequent letters to his close friends, the Nortons.[27]

Wright's first criticism is of Abbot's defense of intellectual intuition or "supersensuous reason" (against Hamilton and Kant) as a distinct faculty of knowledge. Hamilton did not ignore the matter as Abbot presumes, says Wright, but expressly rejected such a faculty of knowledge

as not warranted by an analysis of the mind and its contents; and the difference between his [Hamilton's] doctrine and that of all his opponents rests ultimately in their psychological analyses . . . It seems to me, therefore, that the justness of your criticism of Hamilton turns on the correctness of your preceding psychological analysis of Space and Time as rational ideas, in which you assume a faculty capable of attributing to these relations the negative notions of infinity and incomposite unity. The existence of such a faculty is the only real question between you and Hamilton.[28]

* On the Royce-Abbot affair, see Appendix G.

There is a historical connection between this discussion and Peirce's important attack on immediate knowledge or intuitionism in his "Questions Concerning Certain Faculties Claimed for Man" in the *Journal of Speculative Philosophy*, II (1868). In this article, Peirce, like Wright, interprets Kant's forms of intuition (Space and Time) as forms of "a mental *process*—'the Synthesis der Apprehension in der Anschauung.'" [29] Peirce had been holding long daily discussions with Wright on Kant for nearly two years. Wright did not develop his own empirical theory of the "Evolution of Self-Consciousness" until 1872–73, but both he and Peirce had been set since 1860 on counteracting the neo-Kantian theories of knowledge by a naturalistic theory of the evolution of consciousness and the dependence of knowledge on the use of signs.

The same strictures on the Kantian forms of intuition which prevent us from regarding them when devoid of sensory content as empirical knowledge, Wright said, in defense of his own understanding of Kant against that of Abbot, apply to intuition when called a faculty of self-consciousness or moral knowledge. "The supersensuous reason as a faculty of thought and knowledge is an invention of the rationalists to account for the belief we have in an unconditional, unimaginable reality, but this invention Hamilton regards as unnecessary, and therefore unwarranted." [30]

Two years after this discussion, Abbot, who was now condemned by his fellow churchmen as a "radical," sent Wright a paper entitled "The Radical's Theology," whose contents we can surmise from Wright's criticisms. Abbot had apparently used an argument, similar to that of Chadbourne, which made it necessary to start from some metaphysical absolute but unproved ground of consciousness, just as it is necessary in geometry to start with unproved axioms. Wright's reply is important for the history of logic, as well as for understanding of the extent of his empiricism. Wright maintained, in line with Hume and Mill, that geometric axioms like those of any deductive system of reasoning are demonstrable not deductively but inductively, and because they rest on induction, have only a limited field of application. Hence, no metaphysical appeal to axioms can serve as a logical support for the absolutely universal metaphysical axioms

of theology. No metaphysical theory of consciousness is needed
to account for human or animal or organic consciousness of any
observable form. Abbot's metaphysical theory of consciousness
as a "higher mode of being" and of physical force or uncon-
sciousness as a "lower" mode assumes falsely an absolute sense
of a "higher" and nonexistent opposition between consciousness and force, Love and Fate. Against these transcendental
ideas, Wright's empiricism asserted: "No *real* fate or necessity
is indeed manifested anywhere in the universe,—only a phenomenal regularity." [31]

Science knows nothing about creation, as Abbot's scientific
theology pretends. "What indeed do we know," Wright asks Abbot, "about the creation of organic beings, except perhaps that
they began to exist on the earth within a limited time, and that
various races have succeeded one another? To ascribe to creation
the possession of various powers as means to ends is to put an
anthropological interpretation upon it, is to describe creation as
a manufacture." [32]

Thus Wright, consistent with his defense of British empiricism after his conversion to Darwin, never failed to treat evolution as a scientific *hypothesis* about phenomena within the
limited range of man's observations, affording no logical basis
for anthropocentric metaphysics or theology even when offered
by religious positivists like Abbot. The positivists' ethics and religion of humanity, Wright argued, should be dissociated from
their speculative theism, and from their confusion of the language of morals with that of natural science, owing to their failure to discriminate what Wright is willing to call the "moral and
spiritual facts of religion" from the neutral terms of science.

That 'the mysterious Power of the Universe' really does manifest
mental and spiritual powers at least in men and animals, is a fact which
science fully recognizes; and that these powers are by far the most interesting, to the moral and religious nature of man, of all the phenomena of
nature, is the fact with which religion is chiefly concerned.

I have tried in the above [letter] to sketch briefly the method in which
true science should approach these questions, avoiding as far as possible the
terms which have attached to them *good* and *bad* meanings in place of
scientific distinctness,—terms which have a moral connotation as well as
a scientific one . . . Words have 'reputations' as well as other authorities,
and there is a tyranny in their reputations even more fatal to freedom

of thought. True science deals with nothing but questions of facts,—and in terms, if possible, which shall not determine beforehand how we ought to feel about the facts; for this is one of the most certain and fatal means of corrupting evidence. If the facts are determined, and, as far as may be, free from moral biases, then practical science comes in to determine what, in view of the facts, our feelings and rules of conduct ought to be; but practical science has no inherent postulates any more than speculative science. Its ultimate grounds are the particular goods or ends of human life.[33]

This is as sober and lucid a statement of the objectivity or neutrality of science, of the tentative, relative, pluralistic foundations of both science and ethics, as one can find in American philosophical literature of the nineteenth century.

In his next letter to Abbot, who evidently kept pressing Wright on his phenomenalistic pluralism, Wright conceded a point that Peirce had also urged concerning the inadequacy of the law of association to explain the evolution of knowledge. Wright agreed with Abbot that the mere law of association does not provide an adequate foundation for an experientialist theory of the process of cognition: "This process is not determined solely by the laws of association among the elements of the primitive expressions. There is always an *a priori*, or mnemonic element involved.[34] This functional a priori is a product of the use of signs and habits of memory acquired by one's education and training. Though relations are not intuited by sense without judgment, we must distinguish intuitive from abstract cognition.[35]

To Abbot's question: Is not Idealism "the very negation of objective science?" Wright replied, and repeated the answer later in his notes on Peirce's review of Frazer's edition of Berkeley,[36] that Berkeley's Idealism does not deny the existence of objects, but redefines them as immaterial entities. The important point is that the idealistic metaphysics is as alien to scientific knowledge as any other metaphysic: "There is nothing in positive science, or the study of phenomena and their laws, which Idealism conflicts with. (See Berkeley). Astronomy is just as real a science, as true an account of phenomena and their laws, if phenomena are only mental states, as on any other theory." [37]

Of theistic speculations concerning the existence of God and immortality, no matter how "scientific" they are alleged to be by positivistic theists like Abbot, the verdict of neutral science is, according to Wright, "Not proven." But he hastened to add, "Atheism is speculatively as unfounded as theism, and practically can only spring from bad motives. I mean, of course, dogmatic atheism. A *bigoted* atheist seems to me to be the meanest and narrowest of men. In fact, practical considerations determine that a state of suspended judgment on these themes is the state of stable equilibrium . . . Practical grounds are really the basis of belief in the doctrines of theology." [38]

It was precisely against this state of suspended judgment that William James later urged his "will to believe," condemning Wright's position as "nihilistic" though he agreed with Wright that the main function of religion was the practical one (in the Kantian sense) of supporting ethical character.

Knowing of Abbot's personal predicament in being compelled to abandon his profession as a Unitarian minister, Wright offered the following consolation and advice, based on the view that religious instruction should be primarily ethical rather than metaphysical:

I hope you will be able to continue as a religious instructor, to exemplify how irrelevant metaphysics really are to the clergyman's true influence, —quite as much so, I think, as to that of the scientific thinker. The pursuit of philosophy ought to be a side study. Nothing so much justifies that shameful assumption by ecclesiastical bodies of control over speculative opinions as the inconsiderate preaching of such opinions, in place of the warnings, encouragements, sympathies, and persuasion of the true religious instructor. The lessons which he has to deliver are really very easy to understand, but hard to live up to. To help to live up to the true ideals of life seems to me the noblest, if not the only, duty of the preacher.[39]

The last letter we have from Wright to Abbot was in answer to the latter's request to contribute an article on "The Religious Aspects of Positivism" to a volume to be published by the Free Religious Association in the spring of 1869. Wright declined for reasons that explain why Wright never wrote any book on philosophy, which he regarded as a Socratic art:

A cold thesis, served in a book, does not incite the speculative appetite with me; and I confess to the heartiest sympathy with Plato's preference for a *man*, who can question and answer, rather than for a *book*, which must say much at random, or demand an artist's skill and imagination in the writer. One of the most important of the teacher's or preacher's qualifications, yet one of the rarest, is a knowledge of the hearer's mind, so that his discourse may answer to something, or else raise clear and profitable questions. Most philosophical books, lectures, and sermons seem to me either mechanical performances, or else the offspring of subtle vanity and desire for intellectual sympathy. Let one persuade many, and he becomes confirmed and convinced, and cares for no better evidence . . . The majority of audiences and readers judge by texts and phrases, and apply the touchstone of magical words,—and so think they think.[40]

Experience and the extension of science are better teachers of mankind, Wright concluded, than the rhetorical writings of philosophers: "Men will not agree in the fashions of their dress, in manners, or 'beliefs,' till reduced to the naked facts of experience; and the precepts and methods of modern science, every day extended to new fields of inquiry, will in these, I believe, do more to invigorate and correct the human understanding than all the essays of all the philosophers."[41]

Wright's neutral positivism was not antimetaphysical (as Mach and Stallo in their reactions to neo-Kantianism were), but nonmetaphysical. By this I mean that he regarded science as indifferent to older philosophical speculation in the sense illustrated by Wright in his letters to Abbot. "Why are we Protestants rather than Catholics, Unitarians rather than Orthodox, radicals rather than reactionists? Certainly not for the kind of reason which makes us Newtonians."[42] Wright's argument, if it is still as valid as I think it is, would be ruinous to the contemporary tendency of totalitarian political philosophers to control science as well as religion. Wright qualified his positivism by insisting that scientific philosophy

must be a system of the universal methods, hypotheses, and principles which are founded on them [the growing sciences], and if not a universal science in an absolute sense, yet must be co-extensive with actual knowledge, and exhibit the consilience of the sciences. But while positivism ignores religion in the narrower sense of the word,—that is, the body thereof,—it nevertheless unlike the old atheism, does not reject the religious spirit. It is rather constrained—not for itself, but through the earnest practical characters of many of its disciples—to yield some worthy object to religious devotion, which they think they find in the interests of humanity. But this is an affair of character, not of intelligence. If you define the end of philosophy to be the attainment of religious objects and truths, then positivism is no philosophy. The religion of positivism is no part of its philosophy, but is only a religion which consists with its rigid methods and restraints. Mr. Mill maintains that such a religion is not only possible, but has actually controlled the lives and formed the characters of men of this way of thinking.[43]

Something of the spirit of this religion of character devoted to the interests of humanity, consistent with and yet no part of a rigorous scientific philosophy, appeared to have influenced Abbot and Peirce; however, the former was led by his "Inner Light"[44] to found a new church, and the latter to an apocalyptic vision of evolutionary love, which a knowing historian of American religious thought has assimilated to Unitarianism, because it "was a more sophisticated formulation than Fiske's of the faith in natural progress through the growth of mind."[45]

3. WRIGHT'S DEFENSE OF DARWIN AND SCIENTIFIC NEUTRALITY

The most interesting of the items among the papers of Wright at Northampton is the series of eight letters from Darwin which ended only in the year of Wright's death. The correspondence began when Wright sent Darwin a copy of his long, critical review[46] of an anti-Darwinian book, *The Genesis of Species* (1871), by a Jesuit trained in physiology and law, St. George Mivart. Mivart attacked Darwin's work as "unscientific," and also as morally dangerous for omitting the Divine power as a factor in evolution. His casuistic position is evident in an anonymous review of Darwin's *Descent of Man* which he wrote for the British *Quarterly Review,* concluding with a typically dogmatic declaration that the biological investigations of Darwin (whom he classifies as a "physicist") are erroneous because they are based on a false metaphysic:

Mr. Darwin's errors are mainly due to a radically false metaphysical system in which he seems (like so many other physicists) to become entangled. Without a sound philosophical basis, however, no satisfactory scientific superstructure can ever be reared; and if Mr. Darwin's failure should lead to an increase of philosophic culture on the part of physicists we may therein find some consolation for the injurious effects which his work is likely to produce on too many of our half-educated classes. We sincerely trust Mr. Darwin may yet live to furnish us with another work, which, while enriching physical science, shall not, with needless opposition, set at naught the first principles of both philosophy and religion.[47]

Nothing could have been more opposed to Wright's neutral conception of scientific inquiry than such a declaration, and perhaps it led Wright to emphasize repeatedly in all of his subsequent reviews the importance of preserving scientific questions free from a priori metaphysical partisanship. It was Mivart's scholastic and dialectical method of discussing Darwin, like most queries about the theistic or atheistic character of physical hypotheses (for example, those of Copernicus or Buffon), which Wright regarded as fruitless for further scientific progress, tentative and groping as that progress might be.

Mivart included in his attack on Darwin's work as scientifically unsound the argument that it overlooked the abrupt variations of species manifest in "accidents" and "spontaneous generation." The spontaneous production of life by chemical synthesis out of inorganic elements, Wright argued in his lengthy reply, is just as speculative as Darwin's theory of pangenesis, based on the hypothesis of gemmules or vital molecules; but Darwin's view at least recognizes the empirical continuity between living and inorganic phenomena and the experimental difficulty of reducing living forms to the physical or chemical order of forces, as then known. Thomas Huxley had already attacked Mivart's theological bias by quoting Suarez's views on biology and showing how unfaithful Mivart was to his scholastic authorities. But this was a tour de force, and Darwin, as the correspondence shows, was more impressed by Wright's scientific arguments against Mivart. The latter was a strong adversary because he considered evolutionary biology as essentially a branch of physical science; since "natural selection" claimed to be a law operative in nature but not reducible to the laws of physical science, Mivart could attack it as unscientific without apparently bringing in the real theological grounds of his opposition.

Wright attacked Mivart's sharp and absolute scholastic distinction between "accidents" and "essential" forms of life, which supported the view of abrupt rather than evolutionary variations. Mivart's "accidents," Wright argued, are relative to our ignorance of how determinate causes concur within the limits of what turns out historically and empirically to be useful to the *whole* race. Natural selection depends on such *normal* variations as are useful for group survival, not on the minute chance ones that may have

utility only for an *individual*. No species has absolute constancy
of characters. But some of Wright's arguments, namely, those
based on Lamarckianism, are not as pertinent today as the
methodological principles which he brought to bear heavily and
effectively on the arguments between the scientific view of evolution and its theological interpretations.

To collect together in one place all that can be said for an hypothesis,
and in another all that can be said against it, is at best a clumsy and incon-
venient method of discussion, the natural results of which may best be seen
in the present condition of theological and religious doctrines. These practical considerations are of the utmost importance for the attainment of the
end of scientific pursuit; which is not to arrive at decisions or judgments
that are probably true only on the whole and in the long run; but is the
discovery of the real truths of nature, for which science can afford to wait,
and for which suspended judgments are the soundest substitutes.[48]

Mivart's theistic arguments were shared by no less a scientist than Louis Agassiz, who joined the Catholic adversaries of
Darwin (Mivart, Orestes Brownson, *et al.*) by labeling his views
a temporary "mania," as late as 1867:

My recent studies have made me more adverse than ever to the new
scientific doctrine [of Darwin] . . . This sensational zeal reminds me of
what I experienced as a young man in Germany, when the physio-philoso-
phy of Oken had invaded every center of scientific activity; and yet what is
there left of it? I trust to outlive this mania also.[49]

Methodology is always in the foreground of Wright's critical reviews. In a notice of the sixth edition of Darwin's *Origin
of Species*[50] he suggested that the discussion of rival hypotheses
on the causes of biological evolution might profit from the history of the methods that had been fruitful in astronomy and geology.[51] By that date (1872) the general Darwinian idea of evolutionary change of species had passed beyond the speculative
stage of mere hypothesis, despite the unscientific and rhetorical
opposition of theistic critics:

There have grown up general methods of investigation and discussion
which prescribe limits and precautions for hypothesis and inference; and,
more than all, for the conduct of debate on scientific subjects that have been
of the greatest value to the progress of science . . . These methods are

analogous in their purposes to the general rules in courts of law, and constitute the principles of method in experimental philosophy, or in philosophy founded on the sciences of observation. They serve to protect an investigation from prejudice.[52]

Methodologists like Wright have helped to free scientific method from some of the theological and metaphysical impediments of their times. Even though we cannot credit him with discovering any specific scientific laws, Wright's methodological ideas bore concretely on the scope and limitations of the mathematical-physical sciences, on the promotion of Darwin's work in biology, and on psychology (where "mysticism still reigns" [53]) in the United States, at a time when the teaching of these sciences was struggling to liberate itself from domination by theological and a priori metaphysical thinking.

Wright was not a biologist, but he concentrated his dialectical fire against the reading into Darwin's detailed biological evidence of any metaphysical, theological, or ethical thesis, especially of the sort that came from either spiritualistic or materialistic post-Kantian German philosophy. This was not because Wright was indifferent to metaphysical or ethical questions; all his published and unpublished writings and correspondence are ample evidence to the contrary. His isolation of scientific method from metaphysical speculation and ethical sentiment followed Kant's pluralistic conception of the diverse activities of the human mind. Wright's pluralism, however, was less static than Kant's. He held that the adaptive features of the mind are inseparable from its genetic evolution, in which it has developed from the primary and single function of aiding physical survival to the multiple acquired functions of communication, intellectual exploration, and ethical behavior.

Darwin's reply to Wright's enthusiastic defense of his position was not unnaturally highly complimentary:

Down, Beckenham, Kent
July 14, 1871

My dear Sir,
I have hardly ever in my life read an article which has given me so much satisfaction as the Review which you have been so kind as to send me. I agree to almost everything you say. Your memory must be wonderfully accurate, for you know my works as well as I do myself, and your power of

grasping other men's thoughts is something quite surprising, and this, as far as my experience goes, is a very rare quality. As I read on I perceived how you must have acquired this power, viz., by thoroughly analyzing each word.—I believe Mr. Mivart to be a thoroughly inaccurate man; but he was educated as a lawyer and seems to me to plead, as if retained by a client. —I detected in two places that he gives the commencement of a sentence or paragraph and by omitting the rest attacks my meaning. A Review has just appeared in our Quarterly, evidently by Mivart, and cutting me into mincemeat. It seems to me that you appreciate his kind of intellect with great truth.[54]

The rest of Darwin's letter was a request, which Wright was very happy to grant, that he be permitted to reprint at his own expense Wright's review of Mivart's *Genesis of Species*. In response to this cordial welcome of support, Wright sent Darwin his paper on Phyllotaxy, about which Darwin had inquired in a postscript.[55] Phyllotaxis is an application of elementary mathematics to a theory of the origin and use of the arrangement of leaves around their stems. Wright devoted a great deal of time and writing to it. It is the theme of a five-page article[56] he contributed in 1859 to *Runkle's Mathematical Monthly*,[57] the first mathematical monthly to appear in the United States; Wright was a charter member and sponsor, made several short mathematical contributions, and served on the board of judges to award prizes for the solution of problems.

Phyllotaxy had been suggested speculatively in 1790 by Goethe's essay on the spiral tendency in the development or metamorphosis of plants.[58] It was worked out mathematically in 1834–35 by the German vitalists Karl F. Schimper and his disciple Alexander Braun.[59] These German systematists opposed Darwin's theory of descent (as well as the atomic theory) because it conflicted with their own idealistic *Naturphilosophie*, which subordinated empirical and observable facts to the more satisfying intuitions of the mathematical forms of development from within. When Darwin's work appeared, Schimper expressed himself as follows: "Darwin's doctrine of breeding is, as I discovered at once and could not help perceiving more and more, the most short-sighted possible, most stupidly mean and brutal, much more paltry than the tesselated atoms with which a modern buffoon and hired forger has tried to entertain us."[60] And his more logical and thorough disciple, Braun, who thought he could

reconcile phyllotaxis with Schleiden's discovery of the cell, concluded his vitalistic attack on Darwin: "The inadequacy of the
so-called physical view of nature as compared with the teleological is therefore most felt in the domain of organic nature, where
special purpose in the phenomena of life appears everywhere in
great distinctness." Schleiden himself rejected the vitalistic view
of phyllotaxis as contrary to his own views of plant morphology
and development: "All lines and surfaces which occur in organized bodies are curved, and almost always so irregularly that a
geometrical definition of them is at present out of the question." [61]

After Darwin's work appeared, Wright tried to show that
the spirals and arithmetic formulas (continuous fractions) of
phyllotaxis could be given an empirical and evolutionary interpretation. He discarded the a priori spiritualistic metaphysics of the
German school; and secondly, he combined the genetic characters (overemphasized in the preformationist developmental theories) with adaptive characters due to natural selection. Natural
selection could explain empirically how plants whose leaves are
arranged to ensure the maximum exposure to moisture and sun-
light have a greater chance to survive than plants whose vas-
cular and leaf arrangements are too compact.

Darwin expressed appreciation of Wright's paper on Phyllotaxy, but he was puzzled by the mathematical and the "philosophical" parts: [62]

April 6, 1872

I have read your paper with great interest, both the philosophical and
special parts. I have not been able to understand all the mathematical reasoning; for irrational angles produce a corresponding effect on my mind.
Nevertheless I have been able to follow the general argument, and I am
delighted to have a cloud of darkness largely removed. It is a great thing to
be able to assign reasons why certain angles do not occur, or occur rarely.
I have felt the difficulty of the case for some dozen years, ever since Fal-
coner threw it in my teeth. Your memoir must have been a laborious under-
taking, and I congratulate you on its completion. The illustration taken from
leaves of genetic and adaptive characters seems to me excellent as indeed
are many points in your paper.

You sent me 3 copies, and after reflection, I have sent one to "Nature,"
as one of the editors is a botanist and may notice it; the second I have sent
to the Linnean Society, as most botanists belong to it. I will lend my own
to Mr. Airy (the son of the Astronomer Royal) who has attended to phyl-

lotaxy and who expressed a wish to read your paper. I sent you some time ago my new edition of the Origin which I hope you have received, but pray do not acknowledge it.

Believe me, my dear Sir

Yours very sincerely,

Charles Darwin.

P.S. I have heard that Mr. Mivart will answer, I suppose satirically, your pamphlet in the Popular Science Review, the April number, which ought now to be published. Do you ever see Fraser's Magazine? There is a starting article on Divinity and Darwinism, by I suppose, Y. L. Stevens, who married one of the Miss Thackerays.

In his first letter of July 14, 1871, Darwin also consulted Wright on the application of the theory of natural selection to the evolution of language. The theologians' philological views were rationalistic, and assumed that men had consciously built up useful combinations of sound into words. Darwin here took the view that changes in language occur gradually by what he called "unconscious selection." The great naturalist, who was not given to flattering, says in this same letter to Wright: "As your mind is so clear, and as you consider so carefully the meaning of words, I wish you would take some incidental occasion to consider when a thing may properly be said to be effected by the will of man."

This discussion of "unconscious selection" in the evolution of language was followed less than a year later by Wright's naturalistic account of "The Evolution of Self-Consciousness" in terms of the association of symbols with remembered actions. "Self-consciousness," we must remember, was the pet term of post-Kantian and Hegelian metaphysicians who had little regard and much scorn for an individualistic psychology which did not reflect the "objective mind" of social mores and institutions.

Wright's application of Darwin's views to language and its role in human reasoning, as well as to the natural origins of "self-consciousness," influenced both William James and Charles Peirce. This can be seen by comparing Wright's views, as Perry has,[63] with those of James in the latter's early essay on "Brute and Human Intellect" (1878), and with Peirce's paper in the John Hopkins *Studies in Logic*. Did not James in his psychology accept Wright's insistence on the empirically irreducible differences between animal and human manifestations of intelligence? James's whole philosophical view that the elements of immediate experience are *not* reducible to either the purely mental or the purely physical, his pluralism, in other words, appears in Wright as follows: "Matter and mind co-exist. There are no scientific principles by which either can be determined to be the cause of the other." [64]

The logical principles on which Wright rested his case against both the spiritualistic and the materialistic metaphysics of mind become most clearly explicit in his review of McCosh's views on Tyndall. [65] There are two kinds of explanation of phenomena, one of which refers to the *conditions* of their occurrence and the other of which consists in an *analysis* of effects into their constituents. Mental events, Wright then goes on to say, are fully *conditioned* by physical ones but are not *analyzable* into physical atoms as parts.

That Wright in his attitude to science was metaphysically neutral rather than "antimetaphysical" in a narrow positivistic sense, is clear not only from his criticisms of Auguste Comte, but also from his review of Fraser's *Life and Works of Bishop Berkeley:*

> Science, as such, has nothing to do with faith, but philosophy, which aims to embrace the whole man, can never ignore faith and its data and phenomena. We may try to ignore the knotty problem of "causality," and take things simply as they are, without establishing any doctrine of their relations; we may try to ignore metaphysical and spiritual speculations, and confine our attention to sense-perceptions, but the phenomena remain, and the human mind is certain to revert to them and seek their solution. [66]

In "A Fragment on Cause and Effect" (1873), Wright again attacked both materialism and spiritualism as dubious metaphysical interpretations of science, and later suggested that emotional or value-judgments were responsible for the metaphysics of materialism as well as of spiritualism:

> The metaphysical doctrine of materialism so often charged against or imputed to such scientific thinkers [Carl Vogt, L. Büchner] is, in fact, a doctrine quite foreign to science, quite out of its range. It belongs, so far as it is intelligible, to the sphere of sentiment, moral feeling, and practical principles. A thinker is properly called a materialist when he concludes that his

appetites and passions and actions, having material objects and results for their motives, are those most worthy of serious consideration . . . Scientific doctrines and investigations are exclusively concerned with connections in phenomena which are susceptible of demonstration by inductive observation, and independent of diversities or resemblances in their hidden natures, or of any questions about their metaphysical derivation or dependence. That like produces like, and that an effect must resemble its cause are shallow, scholastic conceptions, hasty blunders of generalization, which science repudiates, and with them it repudiates the scholastic classification or distinction between material and spiritual which depended on these conceptions.[67]

In common language, cause most often means a necessary condition that happens to be prominent or conspicuous. But for Wright, who was not a common-sense realist of the Scottish variety or a Berkeleyan, the experimental method presupposed the universality of causation without metaphysical prepossessions:

The very topic of experimental philosophy, its expectation of constructing the sciences into a true philosophy of nature, is based on the induction, or, if you please, the *a priori* presumption, that physical causation is universal, that the constitution of nature is written in its actual manifestations, and needs only to be deciphered by experimental and inductive research; that it is not a latent invisible writing, to be brought out by the magic of mental anticipation or metaphysical meditation.[68]

The chief historical significance of Wright's defense of Darwin in a hostile theological environment and in an era of German transcendentalism lies in his insistence on the metaphysical and ethical neutrality of scientific method. This concept of scientific neutrality illuminates Wright's support of Darwin in the controversial world of contemporary metaphysical, theological, and ethical opinion.

Wright used the terms "nihilism," "neutrality," and "naturalistic empiricism" synonymously to signify: (1) an emphasis on particular observations and experiments for the piecemeal advancement of human knowledge, without denial of the role of leading questions and hypotheses as guiding principles for research, in accordance with what he frequently called the spirit of "Baconism" and the method of Newton; (2) freedom of the scientific investigator from all forms of extra-scientific control imposed by a priori metaphysical or theological systems or authorities that arbitrarily extend the domain of ethical sentiments to matters of scientific knowledge; (3) the critique of evolution-*ism* as a cosmic generalization, and the defense of evolution in Darwin's presentation as a scientific theory of biology, with no regard for any considerations that might produce unnecessary and unwarranted "conflicts" with religion; (4) the conception of philosophy as a pluralistic scientific methodology rather than as a positive doctrine or peculiar mode of intuitive knowledge.

That scientific inquiry does not logically require any meta-physical, theological, or ethical commitments, Wright regarded as an induction from the history of scientific discoveries. Despite the bitter controversies between antagonistic metaphysical and theo-logical interpreters, the scientific contributions of Copernicus, Galileo, Newton, and Darwin have come to be accepted by scien-tific investigators. The great service of Greek philosophers like Socrates and Plato, Wright remarked in his last year, was in teaching *how* rather than *what* to think and believe. Metaphy-sicians have erected scientific hypotheses, including evolution, into dogmas. Hypotheses are for Wright "trial-questions . . . in-terrogations of nature; they are scaffoldings which must be taken down as they are succeeded by the tests, the verifications of ob-servation and experiment." [69] Even "natural selection" was "to be applied as a working hypothesis in the investigations of general physiology or physical biology." [70]

The method of metaphysics which Wright thought vitiated "German Darwinism" consists in employing methodologically de-tached abstractions, that is, abstractions

without check in definition and precision from the concrete examples, or in treating "realized abstractions" *as if* they had a meaning independently of the things which ought to determine the true limits and precision of their meaning. Thus to apply the mechanical law of the conservation of force, which, as a scientific truth, has no meaning beyond the nature and conditions of material movements (whether these are within or outside of an organism) —to apply this law analogically to all sorts of changes—to the "move-ments" of society, for example—is, in effect, metaphysics, and strips the law of all the merits of truth it has in the minds and judgments of physical philosophers, or of those through whose experimental and mathematical re-searches it came to have the clear, distinct, precise, though technical mean-ing in science which constitutes its only real merits.

Darwin's scientific theory, which despite Mivart and Agassiz, "bears no relation to any 'system' of philosophy," became "Darwinism" in Germany.

Wright thought Darwin's work exemplified "the refinement of modern English Baconism." German historians of philosophy, he points out, have never been able to make out where to place Bacon's so-called philosophy, lacking as it does a "system." No wonder, for it was Francis Bacon who tried to secularize science, make it respectable, and free it from a priori theological and metaphysical systems. Wright credits Bacon with the attitude of scientific neutrality rather than with the discovery of the rules of scientific procedure due to science itself:

> Indeed he [Bacon] had no system [in the German sense]: but by marshalling the forces of criticism known to his time, and reinforced by his own keen invention, against all systems, past and prospective, he aimed at establishing for science a position of neutrality, and at the same time of independent respectability, between the two hostile schools of the Dogmatics and Empiricists, though leaning towards the tenets of theology just so far as these had practical force and value. He thus secured the true status for the advancement of experimental science, or of experimental philosophy, as it came to be called. He had less need of doing, and deserves less credit for what is more commonly credited to him—namely, laying down the rules of scientific pursuit, which the progress of science has itself much more fully determined.[71]

Finally, Wright observes that the same thing has happened historically to Darwin in German metaphysics as had happened to Newton at the hands of his French philosophical disciples.

It would be a historical mistake to believe that Wright was simply a disciple of the positivistic philosophy of either Auguste Comte or of Ernst Mach (1838–1916). Wright of course condemned the theological development of the later stages of Comte's priesthood of scientific sociologists, with its attendant authoritarian hierarchy and abandonment of the neutrality of science.

> In his inability to remain on the neutral ground of positivism, M. Comte lacked one of the virtues of a true philosopher which his critic [Mill] possesses in a marked degree—the power to maintain that commonly painful mental attitude, a suspended judgment. To entertain open questions was

wholly opposed to M. Comte's tendencies as a thinker, and, accordingly, in his later speculations he does not hesitate to substitute a refined kind of fetichism in place of the current theism, not indeed as a true doctrine of science, but as a view of nature which we ought to entertain in our poetical and religious contemplations of it. The barrenness of science, its inability to satisfy the emotional side of our nature, is M. Comte's apology for the religious speculations to which he devoted his second career; but he is betrayed into calling his religious poetical views of nature by the name of "beliefs" thus sacrificing, at least in his language, the fundamental position of positivism.[72]

As to Mach, the only possible reference I have found in Wright is in his review of "German Darwinism," which cites the school of German positivists [perhaps Avenarius (1843–1896), Kirchhoff (1824–1887), Helmholtz (1821–1894)] as an exception to the tendency of German scientists to impose a metaphysical and theological "system" on science:

> There is, however, at present [1875] in Germany an ascetic school of experimental and inductive science, which deprives itself of the aid and guidance of theoretical and deductive considerations in order the more effectively to protect itself from their undue influence. These *Gelehrten* are not true Baconians; but their method might be appropriately named "experimentalism." Men of science in Germany have in general never considered themselves as in a respectable neutral position with respect to opposite systems of philosophy.

Wright's metaphysical neutrality was less emphatic than Mach's more positive "antimetaphysical" position. It was of the British skeptical variety which we find at its highest in David Hume. In the following defense of Hume's skeptical method against critics like McCosh who preferred the method of "common-sense" intuitionism, Wright's naturalistic and evolutionary empiricism suggests that "common sense" and the most apparently secure of "intuitions," whether philosophical, scientific, or "vulgar," should be subjected to scientific criticisms:

> What is sought by the so-called "sceptic" is the nature of the fact, its explanation; and he is not deterred from the inquiry by the seeming simplicity of the fact, but proceeds like the astronomer, and the physicist and the naturalist, by framing and verifying hypotheses to reduce the simple seeming to its simpler reality . . . [The sceptic] denies that the present simplicity of a fact in our thoughts is a test of its primitive simplicity in the

growth of the mind. For such a test would have deterred the astronomer from questioning the Ptolemaic system and the stability of the earth, or the physicist from calling in question nature's abhorrence of a vacuum.

The oracular deliverances of consciousness, even when consulted by the most approved maxims of interrogation, cannot present a fact in the isolated, untheoretical form which criticism and scientific investigation demand. Philosophers are not the only theorizers. The vulgar and the philosopher himself as one of them, have certain theoretical prepossessions, natural explanations and classifications of the phenomena which are habitually brought to their notice—such as the apparent movements of the heavens, and the axioms of hourly experience. How are these natural theories to be eliminated? How unless by criticism—by just such criticisms as those of the great "sceptic" Hume? [73]

No better evidence of Wright's sincerity and steadfast adherence to his method as a scientific critic and philosopher can be given than his abandonment of the Lamarckian idea of "use" as an evolutionary factor, in his last letter to Darwin in February 1875. For "use" implied teleology or purpose in nature, and speculation about ends in nature means abandoning the metaphysical neutrality of science: "The inquiry as to which of several real uses is the one through which Natural Selection has acted . . . has for several years seemed to me a somewhat less important question than it seemed formerly.[74]

Even in defending Darwinism,[75] whose "introduction within the last ten years marks an era of natural science," Wright pointed out that Darwin's speculation concerning pangenesis was no more than a "provisional hypothesis." Although he accepted—as did most of his contemporaries, for example, Samuel Butler, Ernst Haeckel, Herbert Spencer, Walter Bagehot, Alfred Wallace, C. S. Peirce, G. Stanley Hall—the Lamarckian view of the transmission of gradually acquired adaptive characters, Wright was cautious enough to state his acceptance with plenty of "ifs." Alfred Wallace, who, says Wright, anticipated Darwin's *Origin of Species* by a few years in a speculative paper (1855), held to an acceptable form of evolution,

if variations are heritable, in other words, if offspring tend to resemble parents and grandparents, which no breeder doubts and few attentive readers of the present work [Darwin's] will be likely to question; and finally, if the world and the conditions of nature be subject to change, however slow; and the slower the better for natural selection.[76]

Wright finds Wallace's views too "speculative," although it seems to him remarkable that both Wallace (in 1855) and Darwin (in 1859) refer to the same class of positive facts, namely, the distribution of living forms in the Galapagos Islands, as evidence for the same independently discovered hypothesis. Furthermore, both used no theological arguments from design, but found the basis of their hypothesis in (and here Wright fondly quotes from De Morgan) "that exquisite atheism 'the nature of things.'" Wallace went astray, according to Wright, in speculating[77] on the size of the brain as the mark of man's intellectual superiority over his ancestors, overlooking the making, use, and command of artificial signs as a measure of man's greater intellectual power. Thus we see how Wright underscored the function of language in the "origin of consciousness" two years before he extended his naturalistic psychology to the evolution of self-consciousness.

In reviewing Spencer's *Biology* he had judged that popular writer's conception of evolution as "divine agency in creation" to be a piece of a priori cosmology and theology. Spencer had no adequate philosophy of mechanics, Wright asserted, although he was continually speaking of the mechanics or dynamics of evolution. Evolution meant for Wright continuous temporal transmutation of species, but not, as Spencer added, *progress* to an end. "All that the transmutation hypothesis presupposes is continuity and uniformity in the temporal order of nature," Wright observes, whereas in Spencer's theological approach, "evolution expresses more than the evidence warrants." [78] "If we must have a cosmology"—and evidently his friend John Fiske thought so when he presented Spencer's "Unknowable" to the American public with an éclat that aroused Wright's criticism—"if we cannot restrain our speculative faculties to a less ambitious exercise—then that theory of the universe in its totality which agrees best with the facts of science and the ideas with which science has familiarized instructed minds in modern times, is the one most likely to gain credence. But science is itself in no need of such illumination."

Wright applied his conception of metaphysical neutrality in the sharp criticisms of Spencerian evolutionism he included in the last review he wrote, published three days before his death, and entitled "German Darwinism." There he distinguishes books "which treat of evolution as a theorem of natural history from a Baconian or scientific point of view, either mainly or exclusively (confining themselves to considerations of proof), and those which treat of evolution as a philosophical thesis deductively, and as a part of a system of metaphysics." [79] Spencer's ideas of evolution (which were more popular in the United States than in England, thanks to Fiske and Youmans) "were really theological in origin, and have never departed from the theological standpoint." For despite his denial of the existence of a knowable God, Spencer's views reflect a state of mind "in which the old [theological] interest remains and the object still affects the believer as an unconditioned, unproved, undemonstrable, but not less pragmatically real existence; and this is the real starting point of Mr. Spencer's philosophy." [80]

The greatest evidence of Wright's fidelity to his skeptical principle of metaphysical neutrality is his critique of the cosmological generalizations of evolution. Despite his appreciation of Darwin's work, he could not see how it could be used as the basis for a metaphysical doctrine. In his "Physical Theory of the Universe," Wright subjected the evolutionary cosmology to a logical scrutiny unparalleled in rigor in any other work of its time. It was not until seventeen years later that Stallo's searching methodological critique of the cosmological speculations of Kant and others appeared. [81] Wright, like Stallo after him, noted approvingly that Laplace had restricted his form of nebular hypothesis to the solar system, and that, though astronomers no longer held to Laplace's hypothesis, Laplace's evolutionary view still revealed in the *details* concerning the solar system

a physical origin, directing inquiry as to how they were produced, rather than why they exist,—an inquiry into physical rather than final causes; features of the same mixed character of regularity and apparent accident which are seen in the details of geological and biological phenomena; features not sufficiently regular to indicate a simple primary law, either physical or teleological, nor yet sufficiently irregular to show an absence of law and relation in their production. [82]

Instead of speculating like Peirce on the ultimate categories of Law, Chance, and Evolutionary Love, which fused (and confounded) astronomical, geological, biological, and psychological hypotheses, Wright noted critically that while the evolutionary conception of nature was "superior [in generality] to any other generalization in the history of philosophy," [83] it was far from established that the detailed evidence from the various sciences warranted any metaphysical extension of the evolutionary concept from the organic to the vaster inorganic world: "We strongly suspect that the law of 'evolution' will fail to appear in phenomena not connected, either directly or remotely, with the life of the individual organism, of the growth of which this law is an abstract description." [84]

The evolutionary properties of organisms could not be extended to astronomical and geological phenomena except by dubious analogy. Furthermore, Wright wished to confine evolution to natural selection, but here he was subjected to criticism by Peirce.

Aristotle's pluralistic and cyclical view of the ultimate elements and origins of the physical world appeared to Wright to be more sound logically:

Heterodox though the opinion be, we are inclined to accept as the soundest and most catholic assumption, on grounds of scientific method, the too little regarded doctrine of Aristotle, which banishes cosmology from the realm of scientific inquiry, reducing natural phenomena in their cosmical relations to an infinite variety of manifestations (without a discoverable tendency on the whole) of causes and laws which are simple and constant in their ultimate elements. [85]

Scholastic theology, of course, would never have stressed this side of Aristotle, which obviously contradicts the doctrine of creation.

By "physical" explanation Wright did not mean one stated in terms of a "materialistic metaphysic," but simply an antiteleological and neutral account based on hypotheses established inductively and on deductions tested empirically by confronting them with the complexities of observable nature. If materialistic cosmologies hardly explain complexities in taking them as fortuitous, the spiritualistic metaphysics of creation, and of causation by an intelligence which operates with initially predetermined forces, fails to account for the "corrupt mixture of law and apparent accident that the phenomena of the earth's surface ex-

hibit." [86] Wright regarded such irregularities as evidence that our
solar system is a natural product. Laplace's rationalistic materialism erred, according to Wright, in arguing exclusively from the
fact of order: "It is not from the regularities of the solar system,
but from its complexity that its physical origin is justly inferred."

Wright had a nautical title for the operation in nature of
this complexity, contingency, and irregularity, namely, "cosmical
weather." But this term did not signify for him an ultimate metaphysical category of chance operating in Peirce's fashion to create
law and order out of pure chaos by divine statistical prestidigita-
tion. The general fact of complexity and contingency is simply a
perpetual reminder of natural, that is to say, nonteleological causation, and of the fact that scientific theories as well as speculative
generalizations are tentative formulations marked by probable
errors.

Professor Lovejoy has in his critical analysis of the pragmatic
method of defining the meaning of an idea by reference to its
effects or predictable consequences, pointed to a crucial ambiguity: Does the meaning consist wholly in the future consequences predicted by it whether it is believed or not, or in the
future consequences of believing it? [87] It is of some historical interest and importance to note that a similar criticism was made
by Chauncey Wright of John Stuart Mill's and William James's
sensationalism. Wright regarded Mill's definition of material substances as the "permanent possibilities of sensations" to be an important contribution to psychology but not to physics or natural
philosophy. For while Mill is correct in saying we know matter
"only in its effects," he overlooked the distinction made by Hamilton (Wright's early favorite metaphysician), between "effects
on us" and "effects in general." [88] William James, who dedicated
his book on *Pragmatism* to John Stuart Mill, ran into many difficulties by also failing to make this distinction. Stressing the effects
on us of our believing an idea suited James's interest in psychology, but Peirce, in his "pragmaticism," as we shall see in the
next chapter, tried to overcome the subjectivity of James's pragmatism by defining the meaning of an idea as the sum total of all
the conceivable effects which the object of that idea has. But
there are two classes of such general effects: those conceived by
means of verifiable hypotheses, and those more vaguely expressed

by unverifiable metaphysical generalizations. In the important case of the idea of evolution, Wright restricted its meaning to the former class of effects, whereas Peirce was led into an endless labyrinth of metaphysical problems by speculating on a cosmic evolutionism, despite the warnings of his older friend Wright.

4. WRIGHT'S EVOLUTIONARY LIBERALISM AND UTILITARIANISM

Peirce objected to the unethical character of Darwin's doctrines of the "struggle for survival and survival of the fittest" as meaning "every individual for himself and the Devil take the hindmost!" and cited Simon Newcomb's *Principles of Political Economy* as in line with this Darwinian philosophy of "greed." [89] He was thus expressing a moral argument which both Darwin and Wright would say was quite irrelevant to the *biological* issues at stake between, say, Agassiz and Darwin. Wright was a more consistent empiricist and pluralist than Peirce, whose pragmaticism and experimentalism were never reconciled with his own scholastic transcendentalism. Though Wright's philosophy of Darwinian *science* remains neutral with respect to ethical and theological interpretations of evolution, there was no absence of moral feeling in his anxiety to protect scientific inquiry from dogmatic theology. His "nihilistic" neutrality is best understood in the historical context of his impassioned struggle against outworn dogmas that opposed freedom in science, morals, and religion.

There is, therefore, a problem raised by Chauncey Wright's thesis of the neutrality of scientific method; namely, what status did he conceive value judgments to have with respect to the neutral method of science? Are the questions of value inaccessible to our neutral instrument, scientific method, so that we must let unthinking feeling remain in complete custody of our value attitudes? That seems to be implied in the recent positivistic distinction between cognitive and emotive behavior. But for Wright, the relation of thinking to valuing is not one of absolute independence but one subject to cultural evolution. Reflection often leads to new habits which in turn modify old habits and attitudes and thus create new valuations. For example, the original fears and opposition of people to inoculation have been dispelled by education with obvious improvement in public health, although the scientific method that led to the adoption of inoculation is linked with dispassionate studies of the physiology and chemistry of microörganisms. Again, the fact that people cling to unverifiable beliefs about nature, history, politics, or religion calls not merely for the affixing of labels like "meaningless" or "metaphysical" to their beliefs, but signalizes a problem for critical sociological investigation which may lead people to modify their uncritically held attitudes. The neutrality of science did not make ideas concerning economic values inaccessible to analysis; such scientific friends of Wright as Simon Newcomb agreed "that the greatest social want of our age is the introduction of sound thinking on economic subjects among the masses, not only of our own country, but of every other country." [90]

The fact is that Wright did not steadfastly remain neutral toward the social issues of his time. His reviews and letters from 1870 on show less concern about physical science and a gradually increasing interest in the political and moral sciences, including jurisprudence and economics, though not from any desire to defend a system. List's *Political Economy* was espoused by Wright because he realized that the new industries of America, like those of Germany, required protection against British industry. Though he did not favor Mill's radical defense of complete social and political equality for women,[91] he believed their rights as a minority should be protected by a modified form of Hare's electoral scheme of proportional representation.[92] He explained in a long letter to Godkin, editor of the *Nation,* why he agreed with him "on grounds of public policy" rather than with Mill who had assumed the heritability by women of the culturally acquired traits of their educated participation in citizenship. Though Wright regarded sex traits as capable of variable expression in different cultural conditions, there still were unmodified physiological sex differences, especially during gestation, to warrant limiting women's role in government. Political and social evolution called for a gradual extension of their rights.

In economic matters, also, he brought to bear a Darwinian attitude toward class struggle. The whole issue of the status of private property was raised violently in his own time by the revolution of the Paris Commune. Wright's position was not different from the attitude later taken by Justice Holmes toward socialism, namely, that there is nothing inherently wrong in the motive of private possession or accumulation of wealth for one's self or family, but the rights of private ownership are justified not so much by reference to these motives as by the benefits to society of what is produced and distributed even for surplus gain by capitalists. From Wright's rational, utilitarian viewpoint, it is what capital adds to the public store of wealth that makes it deserve the protection of the laws of private property. When these laws fail to perform a publicly useful function and simply protect an unproductive class, they become "devices of legalized robbery, and must be abrogated or amended, if justice is ever to be affected by legislation, through whatever political powers." [93] Now democracy is the most effective modern instrument for solving this problem of the rights of property, even if, as skeptical Chauncey Wright believed, the masses were often ignorant and prejudiced. As he realistically put it to his Cambridge friend, Charles Eliot Norton,

after all, it is a real question, which is the more untoward instrument for the truly just and wise philanthropist to work with,—the ignorant and prejudiced masses whose benefit is sought, or the equally prejudiced aristocracies, blinded by self-interest, whose unjust privileges must be curtailed.[94]

We know that Chauncey Wright was a practical as well as a theoretical utilitarian in political matters. There is an unnoticed letter of his to his senator, Charles Sumner, urging him to continue to fight against the secessionist movement.[95] During the Civil War he asked his youngest brother, Lieutenant Frederick Wright of the Tenth Massachusetts Regiment, to befriend the relatives of a fugitive slave in whose house at Cambridge Chauncey lodged.[96] He held that the so-called "materialistic" aims of the builders of the modern world of business and science were as elevated in moral purpose as the old cathedral builders were supposed to be in the Middle Ages.

Some of the leading spirits of our times are as disinterested and devoted, and find in their aims, whether in politics, industry, or science, as powerful a stimulus to noble passion as the leaders of that age. The masses in all ages are led by the few in all that raises them much above the level of animal wants.[97]

The moral type of modern times who works for the future material well-being of mankind lacks "aesthetic charm," perhaps, but it is "more broadly based on ordinary and universal human interests . . . Shall we on that account call it a lower type?" [98]

There was no conflict in Wright's sensitive and humanitarian mind between the intrinsic ends of beauty and instrumental ends of social utility. The former command our admiration or worship and belong to our aesthetic or religious nature, whereas the latter involve consideration of the practical cost or sacrifice required and belong to our utilitarian moral nature. Wright thus anticipated Santayana's identification of the religious with the aesthetic, but he separated more sharply than Santayana or William James the religious from the ethical, and was thus able to allow a greater diversity of autonomous development for these various sides of human nature.

Moral standards have changed in the course of historical and social evolution, but the relativity of moral ideals to changing historical conditions did not imply their subjectivity. In Wright's judgment, the pragmatic objectivity of morals lies in the fact that ideas generally excite activity in a social milieu. Ideas become dynamically objective "by the quality through which ideas tend to act themselves out." [99] This ideo-motor theory is attributed to Green and Bain ("the grandfather of pragmatism," according to Peirce), and it occurs again and again as the dominant motif of James's dynamic theory of thought as "a fighter for ends," of Peirce's idea of the growth of concrete reasonableness, of Fiske's theory of cosmic progress, and of the more realistic theories of the function of legal thought in Nicholas St. John Green, Joseph B. Warner, and Oliver Wendell Holmes, Jr. Ideas are beliefs on which a man is prepared to act, and thus have an evolutionary or survival value. This biological function of ideas was lacking in the older intuitive philosophies. Wright specifically contrasted his evolutionary conception of thought with the absolutistic view held by post-Kantian idealists in "what they are pleased to call philosophy, that fine composition of poetry under the forms of science, of which Hegelianism is the most notable modern epic." [100] In another of his pithy utterances, he remarked that "philosophy is poetry in the abstract" and "poetry is philosophy in the concrete." Wright was not a narrow utilitarian or my-

opic positivist. He was "the whetstone of wits," Peirce wrote to
James. He also scolded James for having "looked up far too much
to that acute but shallow fellow." Did Peirce mean anything more
than that Wright did not take to his epic of metaphysical evolutionism and pragmaticism?

CHAPTER FOUR

The Evolutionism and Pragmaticism of Peirce*

"Peirce lived when the idea of evolution was uppermost in the mind of his generation. He applied it everywhere," John Dewey has told us in his review of the *Collected Papers of Charles Sanders Peirce*.[1] But Dewey has not told how variously the idea of evolution was interpreted in Peirce's time by Darwinians, Lamarckians, catastrophists, Spencerians, and Hegelians, nor how Peirce's own form of evolutionism was related to his pragmaticism. Peirce never finished any of a dozen or more grand projects intended to be systematic accounts of the logical foundations of the sciences, of the history of thought, and of a scientific metaphysics or cosmology. But the positive results he did achieve are still important for the logic and methodology of the exact sciences, the history of ideas, and philosophy. His many-idea'd speculations, though not as consistent as might have been expected from a rigorous logician, were stimulating enough to have led his lifelong friend, William James, to confer on Peirce the credit of originating pragmatism, albeit through a confusion of names, dates, and ideas.[2] Whereas James saw in evolution the basis of a new psychology, Peirce conceived it as "the cosmic growth of concrete reasonableness," to be understood by means of his own newly furbished logical concepts.

I shall not even attempt to draw a map of the intricate labyrinth of Peirce's philosophy. My more limited aim is to trace in Peirce's intellectual history the role of the idea of "evolution"— or rather, of the several ideas connected in his mind with that term—which may serve as an Ariadne's thread to certain features

* Reprinted, with some changes, from *Journal of the History of Ideas,* vol. VII, no. 3, pp. 321–350 (June 1946).

of his philosophy, especially to his theory of scientific method (including his "pragmaticism") and to his evolutionary cosmology. My plan of exposition has two subdivisions: (1) Peirce's early philosophical conceptions, chronologically or logically prior, in his own mind, to his "evolutionary" logic and cosmology; (2) Peirce's post-Darwinian generalization of biological evolution and his extension of its generalized elements to an evolutionary logic and cosmology (including a novel conception of the evolution of the laws of nature). The conclusions of 1, reached by Peirce before he became preoccupied with the idea of evolution, were the results of his intensive studies in formal logic, semiotic, mathematics, physical sciences (photometry, metrology, and so on), and the critical history of metaphysics, particularly the scholastic realism of Duns Scotus and the transcendental method of Kant; these preëvolutionary studies throw light on Peirce's skeptical attitude toward Darwinian evolution in his logical and cosmological generalizations of evolution to be examined in 2.

1. PEIRCE'S EARLY PHILOSOPHICAL CONCEPTIONS

When Darwin's *Origin of Species* appeared in the United States toward the end of 1859, Peirce was twenty years old, just out of Harvard, and surveying "in the wilds of Louisiana." He was also thinking and writing notes on problems left in his mind by certain discussions at Cambridge with his father Benjamin Peirce, with Henry James, Sr. and his son William, with his "boxing master" Chauncey Wright, and with "Frank" Abbot (Peirce's classmate), about Kant's transcendental logic, especially the question whether it really disposed of a science of cosmology or metaphysics. By correspondence with his Harvard friends, he had heard of the stir that Darwin's work was creating, and of the part played by Chauncey Wright, who held that Darwin's empiricism was in line with the British tradition (Hobbes, Bacon, Locke, Berkeley, Hume, Mill) so dear to Wright and so influential in New England. He remembered vividly the acuteness and unacademic independence of Wright's mind in philosophic conversation among the leading lights of Cambridge, at the Lowells' or the Jameses'. Darwin's *Origin of Species* and Mill's *Examination of the Philosophy of Sir William Hamilton* (1865) had turned Wright from the metaphysics of the Scottish Kantian to the sensationalistic empiricism of the great English liberal; and now Wright saw, in the detailed manner in which Darwin had discovered and demonstrated his theory of natural selection, a vindication of Mill's nonmetaphysical, piecemeal, pluralistic logic. But the logician and metaphysician in Charles Peirce, son of Harvard's most famous mathematician, could not concur with Wright in holding that Darwin's theory of evolution was in line with the sensationalistic nominalism of Mill. Peirce puzzled Wright by arguing that Darwinian evolutionism, because of its positive adherence to "living fact," was bound to destroy the mechanical associationism of Mill's mental chemistry, based on a nominalistic metaphysics rooted in Ockhamism, which Peirce considered the source of the chief errors of modern philosophy.

Now we know approximately what Peirce was thinking about just before and after 1859 from the fortunate fact of his having written at this time, with the exact dates, a score or more of short notes and essays.[3] A glance at the following titles will show why he was bound to be affected by the *Origin of Species* in a way different from Wright.

The first two are college exercises, written when Peirce was eighteen:

May 21, 1857 That the PERFECT is the Great Subject of Metaphysics
Oct. 23, 1857 The Synonyms of the English Language Classified according to their meanings on a definite and stated philosophy[4]

The year of Darwin's *Origin of Species* is one during which Peirce's unpublished notes show no less than fifteen metaphysical essays and memoranda, most of them connected with his close critical study of Kant:

May 21, 1859 That There is no Need of Transcendentalism
May 21, 1859 Proper Domain of Metaphysics
May 21, 1859 New Names and Symbols for Kant's Categories[5]
May 22, 1859 Of the Stages of the Category of Modality or Chance

That Peirce's mind was brought to the idea of chance while studying Kant's categories is noteworthy because of the prime importance this concept later played in Peirce's cosmology and logic of science.

June 1859 Metaphysics as a Study [48 pp.]

July 25, 1859 Comparison of our Knowledge of God and of Other Substances

July 25, 1859 Proposition. All unthought is thought of

July 25, 1859 Of Realism: It is not that Realism is false; but only that the Realists did not advance in the spirit of the scientific age. Certainly *our ideas are as real as our sensations.* We talk of an unrealized idea. That idea has an existence as noumenon in our minds as certainly as its realization as such an existence out of our mind. They are in the same case. *An idea I define* to be the noumenon of a conception.

[On the other side of this sheet, Peirce wrote]: List of Horrid Things I am: Realist, Materialist, Transcendentalist, Idealist.

July 27, 1859 Sir William Hamilton's Theory of the Infinite

Oct. 23, 1859 Two Kinds of Thinking

Oct. 23, 1859 That we can Understand the Definition of Infinity

Oct. 25, 1859 That Infinity is an Unconscious Idea

Oct. 25, 1859 Why We Can Reason on the Infinite

Oct. 25, 1859 Of Pantheism: What is the portion of the soul of which we are not conscious? How does it differ from other men's souls? If consciousness is not our limit, we have no limits but are agglomerated. Another man's consciousness however is not part of me. Neither is noumenon. This is *formal* pantheism.

This idea of a formal agglomeration of minds was later developed (in 1868) into the conception of an ideal community of minds whose investigations are destined to lead to truth. "Logic is rooted in the social principle." [6]

Oct. 25, 1859 Of Objects: 1. Thing 2. Influx 3. The Unconscious[7] Idea 4. Act of Thinking 5. The Soul

Since Peirce tells us that he did not return from Louisiana until early in the summer of 1860, we may assume that he was too busy surveying in Louisiana to do any metaphysical writing for about six months after Darwin's work appeared in the United States in November 1859.

May 30, 1860 Metaphysical Axioms and Syllogisms

June 30, 1860 The Fundamental Distinction of Metaphysics: Thought and the thought-of or the thinkable and the thinkable-of

July 1, 1860 The Keystone of this System

July 3, 1860 The Infinite, the Type of the Perfect

July 3, 1860 *The Logical and the Psychological Treatment of Meta-*

physics: Two methods of viewing metaphysics give rise to two methods of treating it. One starts by drawing the conceptions from logical relations° and thence reasoning to their place in the mind; the other starts by drawing the conceptions from the system of psychology and reasoning to their logical meaning. The former seems to me, if less psychologically exact, to be more metaphysically true in its results, and it is the method I adopt.

This important methodological note is an early anticipation of Peirce's objective or logical realism, which he identified historically with Duns Scotus in opposition to the nominalism and psychologism of Ockham, Locke, Berkeley, Hume, Mill, and Wright.

———, 1860 The Rules of Logic Logically Deduced

July 13, 1860 The Orders of Mathematical Infinity

Aug. 6, 1860 Introductory to Metaphysics (1st draft, 4 pp.)

Aug. 11, 1860 Same (2nd draft)

———, 1860 I, It, Thou

———, 1860 On Positivism

———, 1860 Kant

Aug. 21, 1861 Book on "Metaphysical and Private": Sec. 1 On the Definition of Metaphysics: I believe in mooring our words by certain applications and letting them change their meaning as our conceptions of the things to which we have applied them progresses [page 1] . . . That meditation which gives us new conceptions is a *cultivation* resulting in a growth of thoughts, and the result of the growth of the mind as displayed in the thoughts is called Wisdom. . . . Why is metaphysics so hard to read? Because it cannot be put into books. You may put suggestions towards it into books but each mind must evolve it for himself and everyman must be his own metaphysician . . . The real worth of Metaphysics must lie, of course, in its practical application [page 12] . . . To learn how to analyze ideas, therefore, and to analyze them—in short to study metaphysics—will be *par excellence* education [page 17].

Aug. 21, 1861 Another Attempt at Metaphysics

Mar. 30, 1862 Metaphysics—*Odi profanum vulgus*—In our investigations, metaphysics is to be taken as the analysis of conceptions.

June 8, 1862 I now make out the following as irreducible conceptions: I, Cardinal, Secondary, Sensation, Conception, Abstraction, Metaphysical, Dynamical, Mathematical, Physical (10 in all). These seem to form by their combinations 216 others.

Aug. 5, 1864 A Treatise of the Major Premises of the Science of

---

° "The little I have contributed to pragmatism (or, for that matter, to any other department of philosophy), has been entirely the fruit of this outgrowth from formal logic, and is worth much more than the small sum total of the rest of my work, as time will show" (*C.P.*, 5.469). But how can formal logic lead to metaphysical evolutionism?

<Nature> Finite Subjects: I will not spend one word to prove that this *Metaphysics* is theoretically essential to science, for it is not needed. But since it is generally thought to be an idle study, I will undertake to show that some of the greatest controversies of our day are at bottom questions of the validity of these assumptions.

In the winter of 1866–67, Peirce gave a series of Lowell Lectures on "The Logic of Science and Induction" which James attended * but "did not understand." [8] Yet a few years later he heard Peirce enunciate the principle of pragmatism. Now, it is quite unlikely that James could have understood Peirce's principle when he first offered it, since it was for Peirce primarily logical, intended to clear up metaphysical issues, such as those that bothered him in the years he had devoted to Kant and Hamilton. About the time Peirce says he discoursed on the fixation of belief and how to make our ideas clear to the Metaphysical Club, he was engaged on his Harvard "Lectures on British Logicians" (1869)[9] and "Notes for Lectures on Logic" (1870), as well as on an elaborate review of Fraser's edition of Berkeley for the *North American Review* (1871).[10] In the latter review Peirce struck hard at nominalism and declared for Scotist realism or the objective reality of universals *in re* as the indispensable presupposition of all science. Chauncey Wright praised Peirce's review in two notes in the *Nation*,[11] but of course qualified his praise of Peirce's erudition with reservations about the necessity for dropping nominalism altogether as a presupposition of science. Peirce retorted less politely with a sarcastic "Letter to the Editor" [12] about the inability of contemporaries to understand the true basis of science. No wonder, then, that Peirce and Wright did not see eye to eye on the significance of Darwin's work. Peirce argued persistently that nominalists like Hume, Mill, and Wright have substituted the uniformity of nature for God as an ultimate fact.

When asked by the Secretary of the Smithsonian Institution, S. P. Langley, in 1901, to write a popular essay on "the idea of the law of nature among the contemporaries of David Hume and among advanced thinkers of the present day," Peirce thought it necessary to trace Hume's idea of laws of nature to the fourteenth-century nominalism of the heretic Ockham, and to maintain that the generality of educated men still entertained

* Oliver Wendell Holmes, Jr. also attended some of these lectures.

the same Ockhamistic conception which was commonest in Hume's time; for most men whom I meet, when they refer to such matters [as the laws of nature], talk the language of Mill's *Logic*. In particular the explanation of prognosis most common is that it is rendered possible by the uniformity of nature, which is an "ultimate fact." This adapts itself well to the atheistic opinion which has always been common among Ockhamists,—more so, perhaps, about 1870 than at any other time.[13]

What people like Mill and Wright did not and still do not realize is the hostility between the atheistic individualism of Ockham and the idea of evolution, uppermost in most men's minds at the end of the nineteenth century, Peirce observed.

"Nominalism" is broadly and recurrently used by Peirce to designate several historically affiliated but logically distinct positions: (a) the denial of the existence *in re* of universals or abstract ideas except as general terms, signs or vocables (Roscellinus, William of Ockham, Hobbes, Locke, Mach); (b) the view that all general concepts and laws exist exclusively in the divine or human mind (Abelard, Leibnitz, Berkeley, Kant, Hamilton, Hegel, Comte, Pearson); (c) the restriction of all human values to the individual feelings and desires of men (Hume, Mill, Wright, James); (d) the denial of any universal purposive laws or goals of evolution in cosmic or human history.

When Peirce says that in his important review in 1871 of Fraser's edition of Berkeley, he "declared for realism," it becomes clear, in the light of his other writings at about that time (1870–1875), that in his continued attacks on nominalism, he was aiming to refute Wright's defense of the agnosticism of Mill and Darwin and of the ethical and theological neutrality of science, as set forth later in Wright's critique of the neo-Kantian speculations of "German Darwinism." [14]

If Wright saw in Darwin's work an exemplary scientific product of Mill's empiricism, and eschewed all metaphysical evolutionisms, Peirce, on the other hand, questioned even the *scientific* validity of Darwin's theory, and while admitting the empiricism of Darwin's method, denied that it was in line with Mill's empiricism (which was for him "pure metaphysics"). Peirce's own preoccupation with medieval realism and neo-Kantian idealism led him later, in 1892, to transmute evolution into a cosmology of "evolutionary love," which for sheer speculative audacity was a worthy rival of the absolutism of Schelling and Hegel.

It is not sufficiently recognized that Peirce was less than lukewarm toward Darwin's theory of natural selection as a scientific hypothesis. His logical spirit was no doubt ardent about the possibility of generalizing certain elements of the pre-Darwinian as well as the Darwinian conceptions of organic evolution into an evolutionary philosophy that would rival Aristotle's[15] in its aim of staking out the future lines of growth of all the sciences. By substantiating each of these statements from the texts of Peirce, we shall obtain a clearer picture of the peculiar admixture of the scientific and the metaphysical (or theological) strains in Peirce's evolutionism. We shall thus see why Peirce's evolutionism was not simply Darwin's hypothesis of Natural Selection, but a certain deliberate generalization of it in Peirce's own speculative form.

In the first place, Peirce regarded Darwin's view as indicating only one of three equally operative modes of the evolution of organic species: (1) Darwin's "successively purely fortuitous and insensible variations *in reproduction*"; (2) Lamarck's mode of the inheritance of acquired characters, which assumes continuous, very minute changes due wholly to strivings or efforts of individuals in adapting themselves to the environment; (3) Cuvier's and Agassiz' defense of the cataclysmal mode of large abrupt adaptive changes in reproduction.[16] All three of these modes of evolution have been operative, according to Peirce, and (despite Weismann's poorly contrived experiments), "it is probable that the last has been most efficient."[17] Furthermore, "it is not the sublimity of Darwin's theories which makes him admired by men of science, but it is rather his minute, systematic, extensive, strict, scientific researches which have given his theories a more favorable reception—theories which in themselves would barely command scientific respect."[18] Darwin's view does not apply to man's "unpractical" interests:

> Logicality in regard to practical matters . . . is the most useful quality an animal can possess, and might, therefore, result from the action of natural selection; but outside of these . . . upon unpractical subjects, natural selection might occasion a fallacious tendency of thought.[19]

As late as 1893, Peirce still regarded Darwin's theory as unworthy of much scientific respect:

What I mean is that his hypothesis, while without dispute one of the most ingenious and pretty ever devised, and while argued with a wealth of knowledge, a strength of logic, a charm of rhetoric, and above all with a certain magnetic genuineness that was almost irresistible, did not appear, at first, at all near to being proved; and to a sober mind its case looks less hopeful now [1893] than it did twenty years ago; but the extraordinarily favorable reception it met with was plainly owing, in large measure, to its ideas being those toward which the age was favorably disposed, especially, because of the encouragement it gave to the greed-philosophy.[20]

Wright would certainly have said, in the name of the ethical neutrality of science, that the "greed-philosophy" of those "social Darwinists" who tried to justify rugged and ruthless economic exploitation, was in no wise relevant to the scientific validity of Darwin's hypothesis.

Peirce's preference for the Lamarckian and cataclysmic views was based not on the scientific evidence of biology, but on the neat ways in which they fitted into his metaphysical and theistic evolutionism. In order to understand historically Peirce's attitude to Darwin's work, the influence of Louis Agassiz (d. 1873) on Peirce should be taken into account. Agassiz' *Essay on Classification*, in the first volume of his *Contributions to the Natural History of the United States*, which regarded the forms and orders of all species as eternal thoughts of the Creator, owing their survival or extinction to His Plan, Peirce greatly admired and adopted as the basis of his own *Classification of the Sciences*.[21] It is in a footnote to the latter that we find Peirce admitting that he was influenced by Agassiz in defending the objective reality of natural classes as genuine ideas whose existence depended on a structural hierarchy of final causes: "I am here influenced by the *Essay on Classification* (1857) of L. Agassiz, whose pupil I was for a few months. This work appeared at a most inauspicious epoch." [22]

The fact that Agassiz continued to attack Darwinism in biology[23] even after its acceptance by such an eminent botanist as Asa Gray, friend and teacher of Chauncey Wright, did not deter Peirce from saying as late as 1910, "Agassiz, in his *Essay on Classification*, described well—I do not say perfectly, but relatively, well—what a classification of animals ought to be." [24]

Peirce admits that subsequent zoologists found that Agassiz' classification "did not seem to be a good fit to the facts of the animal kingdom," and then virtually gives both Agassiz' and his own case away by asking, "What wonder? It required the taxonomist to say what the idea of the Creator was, and the different manners in which the one idea was designed to be carried out. How can a creature so place himself at the point of view of his Creator?" [25] However, by the time Peirce reached the end of the metaphysical spiral of his own evolutionism, he came fairly close to adopting the point of view of the Creator of this and of all possible universes.

The historical order of Peirce's writings, lost in the edited order but recoverable from the unedited and unpublished papers, shows plainly that traditional metaphysical and logical problems and ways of thinking not only antedated his evolutionism but also that he fought off the Darwinism of his times and never accepted Darwin's theory of natural selection as sufficient to explain organic or intellectual evolution. We must take into consideration Peirce's own statement that he had learned little from the evolutionary philosophers.[26] Since it is sound methodological principle in intellectual history *not* to assume the infallibility of a thinker's own memory, I have checked Peirce's statements of the history of his own thought by examining the topics of the papers and projects he was busy at during the years 1859 to 1893, the last being the time when the idea of evolutionism begins to occur most frequently in his "Lectures on the History of Science."

Scant justice has been done to Peirce as a historian of ideas in the meager assortment of pages selected by the editors of his *Collected Papers*. These contributions are not merely notes but include carefully prepared and delivered lectures, mounting to hundreds of manuscript pages—for example, twelve Lowell lectures (1893) on the History of Science from the time of Ancient Egypt and Babylonia to Copernicus and Newton, a history of medieval logic (a course given by Peirce at Johns Hopkins as Lecturer in Logic [1879–1884]), plans for an edition of Petrus Peregrinus' work *On the Lodestone*, nine lectures on British Logicians, "The Idea of a Law of Nature Among the Contemporaries of David Hume and Among Advanced Thinkers of the Present Day" (title suggested by the letter of the secretary of the Smithsonian Institution), Studies in Nineteenth-Century Men of

Science, and Experiments for the Study of Comparative Biography (in his logic courses at Johns Hopkins). Peirce's main interest in all these historical studies was the logic of methods of inquiry in all branches of thought. Both he and Chauncey Wright were great admirers of Whewell's *Philosophy and History of the Inductive Sciences.*[27]

What Wright and Peirce in 1856 admired in Whewell's work was its union of Kantian and British empiricist conceptions of scientific laws as summary and cumulative "colligations" of observations. Wright had more than once insisted that the methodology of the empirical sciences should be derived from the cumulative progress of the methods by which scientific discoveries like Newton's had been made. Kant had suggested, in his doctrine of method in the *Critique of Pure Reason,* a study of the history of science that would illustrate the use of the forms and categories posited in his architectonic of the mind. But Peirce, probably under the influence of the evolutionary conception of consciousness which Wright had elaborated in his "psycho-zoölogy" or "Evolution of Self-Consciousness" (1873) in response to a question of Darwin's, "When may a thing be properly said to be effected by the will of man?" applied the evolutionary approach to the whole field of intellectual history:

> The evolutionary theory in general throws great light upon history and especially upon the history of science—both its public history and the account of its development in an individual intellect. As great a light is thrown upon the theory of evolution in general by the evolution of history, especially that of science—whether public or private.[28]

Peirce himself tells us, forty years after Darwin's *Origin of Species* had appeared, that during all the intervening years he had been constantly occupied with the study of methods of inquiry, and that the history of thought provided him with the materials for that study. The unpublished materials fully justify Peirce's statement. While the theory of signs and the logic of relations were the chief products of his earliest methodological investigations, his later contributions to inductive logic and to the history and philosophy of the empirical sciences are more closely related to the impact of evolutionism, until he finally maintained that "philosophy requires thorough-going evolutionism or none." [29]

2. PEIRCE'S POST-DARWINIAN EVOLUTIONISM

Now in order to make the broadest possible use of the idea of evolution, to extend it to encompass the history and logic of thought—despite Wright's warnings against all speculative extensions of evolution, which had, after all, been thus far a working hypothesis for biologists only—Peirce enlarged the idea of evolution to include not only a generalized form of Darwin's view but also two others, Lamarck's and the cataclysmic view.

Wishing to keep his metaphysical theories rooted in the best scientific views of his day, he first dissected the hypothesis of Darwin into its logical elements and then generalized in quasi-mathematical fashion. In Peirce's analysis, the logical elements of Darwin's theory consisted of (1) accidental variations *in reproduction*, (2) elimination of breeds whose birth-rate drops too far below their death-rate in the struggle for survival, (3) the preservation of those species whose characters are favorable to their continued existence ("survival of the fittest"). Now comes Peirce's mathematical generalization of Darwin's view into a theorem in the logic of probability:

> This Darwinian principle is plainly capable of great generalization. Wherever there are large numbers of objects having a tendency to retain certain characters unaltered, this tendency, however, not being absolute but giving room for chance variations . . . there will be a gradual tendency to change in directions of departure from them.[30]

Another more recondite form of Peirce's generalization of Darwinism was given later by him in the following passage:

> In biology, that tremendous upheaval caused in 1860 by Darwin's theory of fortuitous variations was but the consequence of a theorem in probabilities, namely, the theorem that if very many similar things are subject to very many slight fortuitous variations, as much in one direction as in the opposite direction, which when they aggregate a sufficient effect upon any one of those things in one direction must eliminate it from nature, while there is no corresponding effect of an aggregate of variations in other reaction, the result must, in the long run, be to produce a change of the average characters of the class of things in the latter direction.[31]

After then attempting to deduce by means of this statistical cosmologic theorem that the number of surviving or adaptive variations must increase, Peirce adds:

> Anyone who is old enough, as I am, to have been acquainted with the spirit and habits of science before 1860, must admit that in this case, at any rate, the work of elevating the character of science that has been achieved by a simple principle of probability has been truly stupendous.[32]

In 1881 at Johns Hopkins, advanced students of logic were discussing Peirce's theory of induction, "founded upon the material view of probabilities and the theory of the adaptation of the mind to the universe," as we read in an account of the May meeting that year of the Metaphysical Club of Johns Hopkins University.[33]

The statistical generalization that accounts for changes by the operations of chance is illustrated by Peirce from the gambling saloon (where, historically, the mathematical theory of probability originated, thanks to the answers of Fermat and Pascal to the queries of gamblers on betting odds). Suppose, Peirce's illustration runs,[34] a million gamblers, each with one dollar as his sole fortune, play against a bank by betting a dollar on odd and even. After the first trial, about half will have lost their fortune and been eliminated, but the remaining 500,000 will have won and doubled their wealth. After a second trial, about 250,000 will be set back to their original fortune of one dollar, but the other 250,000 will have three dollars. Now, a third trial will eliminate about half of those who had only one dollar, will leave about 250,000 with two dollars and 125,000 with four dollars. And so as the trials increase, the population gets smaller and smaller, but more prosperous. It is remarkable that Peirce took his mathematical analogy seriously as an illustration of his metaphysical generalization of Darwin's theory.

However, there is the notion of chance variation in Darwin's idea of evolution; and Peirce, so far as I know, was the first to show the historical connection of this idea of chance in Darwin with other uses of the same statistical concept in other departments of nineteenth-century science, which were progressing by unprecedented leaps and bounds, namely, statistical mechanics and statistical sociology. From the standpoint of Peirce's interest in the logic of probability, the whole Darwinian controversy was not primarily the occasion for a new theological dispute, or a new revolution in morals or education:

> The Darwinian controversy is, in large part, a question of logic. Mr. Darwin proposed to apply the statistical method to biology. The same thing has been done in a widely different branch of science, the theory of gases. Though unable to say what the movements of any particular molecule of gas would be on a certain hypothesis regarding the constitution of this class of bodies, Clausius and Maxwell were yet able, eight years before the publication of Darwin's immortal work, by the application of the doctrine of probabilities, to predict that in the long run such and such a proportion of the molecules would, under given circumstances, acquire such and such velocities . . . and from these propositions were able to deduce certain properties of gases, especially in regard to their heat-relations. In like manner, Darwin, while unable to say what the operation of variation and natural selection in any individual case will be, demonstrates that in the long run they will, or [would], adapt animals to their circumstances. Whether or not existing animal forms are due to such action, or what position the theory ought to take, forms the subject of a discussion in which questions of fact and questions of logic are curiously interlaced.[35]

But Peirce in his statistical conception of law went much farther than Darwin in biology, Maxwell in physics,[36] or Quetelet in sociology. The latter three scientists, in keeping with an established philosophy of mechanical determinism, regarded chance as no explanation at all but a makeshift concept to patch up our ignorance of more fundamental mechanical or dynamical laws governing every individual event, thing, or character with strict necessity. It is in its moral opposition to this mechanistic or necessitarian assumption in both scientists and philosophers from Ockham's time to the nineteenth century (Hegel and Spencer) that Peirce's philosophy of evolution may be historically understood. He was not content with Wright's attitude of scientific neutrality with respect to the metaphysical and religious disputes over the works of a Newton or a Darwin. Science was a social enterprise for Peirce, and a matter of life and death to those whose minds were nourished by its problems and devoted religiously to its cause, Truth. The flaming torch of science had to spell out in fiery letters: DO NOT BLOCK THE ROAD TO INQUIRY. The "seminary-type" of metaphysics and theology was an age-old obstacle to the growth of the laboratory-inspired scientific philosophy of the experimental mind, and if one did not attempt to construct the true system of metaphysics within which the sciences would receive protection from the fatal chill of lifeless and paralyzing antireligious positivistic philosophies, the sciences would be unable to grow and prosper. Yes, one should not decline, as Wright's "nihilism" did, to help build the temple of the religion of science.[37]

The sort of metaphysics which ought to be built

to represent the state of knowledge to which the nineteenth century has brought us . . . would be a Cosmogonic Philosophy. It would suppose that in the beginning—infinitely remote—there was a chaos of unpersonalized feeling, which being without connection or regularity would properly be without existence. This feeling, sporting here and there in pure arbitrariness, would have started the germ of a generalizing tendency. Its other sportings would be evanescent, but this would have a growing virtue. Thus, the tendency to habit would be started; and from this, with the other principles of evolution, all the regularities of the universe would be evolved. At any time, however, an element of pure chance survives and will remain until the world becomes an absolutely perfect, rational, and symmetrical system, in which mind is at last crystallized in the infinitely distant future.[38]

Here is a chapter from the Book of Genesis for the religion of science after Darwin. Yet Peirce, in one place, regarded his evolutionist philosophy as no more than

such conjecture as to the constitution of the universe as the methods of science may permit, with the aid of all that has been done by previous philosophers. I shall support my propositions by such arguments as I can. Demonstrative proof is not to be thought of. The demonstrations of the metaphysicians are all moonshine. The best that can be done is to supply a hypothesis, not devoid of all likelihood, in the general line of growth of scientific ideas, and capable of being verified or refuted by future observers.[39]

The *raison d'être* of Peirce's pragmaticism, he tells us, is that "it will serve to show that almost every proposition of ontological metaphysics is either meaningless gibberish . . . or else is downright absurd." [40] One wonders, therefore, what would happen to much of Peirce's evolutionistic speculations if he had adhered to the "prope-positivism" of his view that, in consequence of his pragmaticism, "what will remain of philosophy will be a series of problems capable of investigation by the observational methods of the true sciences." [41]

Yet it is clear from Peirce's unpublished essay "On Positivism" [42] that he is not a positivist in the sense current in his time or in our own. He wished to purify philosophy of antiscientific methods and direct it toward certain religious and moral ends. This is evidenced by his statement of what distinguishes his own prope-positivism from other species of positivism. The latter conceal an impure metaphysics, do violence to the instinctive beliefs of mankind, and do not appreciate the truth of scholastic realism or the close approximation to it in F. E. Abbot's *Scientific Theism*. [43]

It is not easy to see any but a far-flung analogy between the elements of Peirce's metaphysics and those of Darwin's theory, which would place the former "in the general line of growth" of the latter. It is difficult to understand how some of Peirce's cosmogonic conjectures are capable of verification. How verify a nonexistent "chaos of unpersonalized feeling" by any procedure which the methods of science may permit? In any case, Peirce regarded Darwin's "chance variations" as illustrating his *metaphysical* category of Firstness, Spontaneity, Variety, Contingency; and he regarded Darwin's "struggle for existence" as illustrating Secondness, Resistance, Force, Brute Existence. Finally, in Peirce's evolutionistic metaphysics the quasi-teleological idea of natural selection by survival of the fittest and the Lamarckian reproduction of adaptive characters, illustrate Thirdness, Law, Purpose, Habit, Generality. The function of intellect becomes primarily one of adaptation to a universe growing in reasonableness.

By the "growth of reasonableness" Peirce evidently meant the gradual predominance of a growing variety of stable habits over random or chance variations, in both the physical universe at large and in the life of the human mind. Such a gradual ascendency of order and habit over the indeterminateness and brute impulses from which they emerged seemed to Peirce to be evidenced by the evolution of both the physical world and that of human thought and civilization. In order to unify physical, biological, psychical, and social evolution, Peirce resorted to a pan-psychical cosmology in his doctrine of evolutionary love (*aga-pism*), in which the First Principle or Category of Firstness is Spontaneous Feeling. "Objective chance"—in one of the various meanings Peirce gives to it in his doctrine of *tychism*—is the un-eliminable and inexplicable variety of qualities that constitute the "first" general feature of experience. The brute, dynamic exist-ence of qualities, events, things constitutes Secondness. But the individual thisness and externality of qualities, things, or events do not exhaust reality in Peirce's metaphysical scheme. There is the Third Principle of General Law, Habit, Purpose, or Conti-nuity (*synechism*). It is this "third" which asserts the objective reality of a variety of evolving laws or habits; they coalesce feel-ings or ideas in the struggle for existence, thus giving the evolu-tionary process a goal. This goal, so far as I can understand Peirce's "evolutionary love," oddly adumbrated his Platonic love of a rich diversity of evolving laws not fully realized in the tem-porally variable laws of the most exact sciences.

> Now variety is the most obtrusive character of the universe of Nature. Was all this variety introduced in one dose in the beginning or has it been gradually developed? All that we have been learning of Nature since the Origin of Species convinces us that this variety was evolved . . . Hence, to say that the variety of Nature grows is to say that events do not exactly conform to any law. There must be slight chance departures analogous to, though of course millions of times smaller than, our errors of observations.[44]

In such passages as this, Peirce posited an increasing variety and randomness in the processes of evolution. But in his view of evolution as a generalized theorem of probability, he emphasized the gradual elimination of chance variations as a necessary step in the growth of cosmic reasonableness. These two views are in-consistent, as Professor Lovejoy indicates.

Peirce applied his evolutionism to three fields which he claimed both supported and were supported by his doctrine: (1) intellectual history, especially the history of science; (2) the logic of the sciences,[45] especially that of probable induction; (3) the metaphysics of history and of science. The fact, however, that Peirce has made important contributions to these fields does not prove that his evolutionism was chronologically prior to or causally responsible for these contributions, as may be learnt from the development of his thought.

The logical studies of Peirce after 1860 from which he derived his "pragmaticism" show an increasing regard for the temporalistic, evolutionary, and experimental features of thought. In every one of the three essential types of scientific inference, Hypothesis, Deduction, and Induction, evolutionary considerations continually crop up in Peirce's extensive and path-breaking logical studies. Hypothesis is a weak quasi-fortuitous surmise in the first tentative gropings of the mind toward the solution of a problem. A problem itself is occasioned by a biological irritation or a stimulus of doubt, the removal of which constitutes the "fixation of belief" and the purpose of thinking. The logic of hypothesis and verification is unintelligible apart from such purposive, "evolutionistic" functions of thought, directed experimentally toward the future settlement of present doubts. Thoughts are subject to the same organic evolution as living species.

To say that the future does not influence the present is untenable doctrine. It is as much as to say that there are no final causes, or ends. The organic world is full of refutations of that position. Such action [by final causation] constitutes evolution . . .

All our knowledge of the laws of nature is analogous to knowledge of the future, inasmuch as there is no direct way in which the laws can become known to us. We here proceed by experimentation. That is to say, we guess out the laws bit by bit. We ask, What if we were to vary our procedure a little? Would the result be the same? We try it. If we are on the wrong track, an emphatic negative soon gets put upon the guess, and so our conceptions gradually get nearer and nearer right. The improvements of our inventions are made in the same manner. The theory of natural selection is that nature proceeds by similar experimentation to adapt a stock of animals or plants precisely to its environment, and to keep it in adaptation to the slowly changing environment. But every such procedure, whether it be that of the human mind or that of the organic species, supposes that effects will follow causes on a principle to which the guesses shall have some degree of analogy, and a principle not changing too rapidly. In the case of natural selection, if it takes a dozen generations to sufficiently adapt a stock to a given change of the environment, this change must not take place more rapidly, or the stock will be extirpated instead of being adapted. It is no light question how it is that stock in some degree out of adjustment to its environment immediately begins to sport, and that not wildly but in

ways having some sort of relation to the change needed. Still more remarkable is the fact that a man before whom a scientific problem is placed immediately begins to make guesses, not wildly remote from the true guess.[46]

Even the deductive reasoning of mathematics bears the
marks of a certain kind of exploratory and ideal experimentation
with diagrams and symbols:

A satisfactory evolutionary logic of mathematics remains a desideratum
. . . It has come about through the agencies of development that man is
endowed with intelligence of such a nature that he can by ideal experiments ascertain that in a certain universe of logical possibility certain combinations occur while others do not occur. Of those which occur in the ideal
world some do and some do not occur in the real world; but all that occur
in the real world also occur in the ideal world. For the real world is the
world of sensible experience, and it is a part of the process of sensible experiences to locate its facts in the world of ideas.[47]

The location of sensible facts in the ideal world requires a
different kind of experimentation from the purely manipulative
sort, but it is significant that Peirce wishes to extend the process
of experiment to deductive reasoning. This extension is understandable as part of the evolutionary notion that Wright had developed in his "Evolution of Self-Consciousness," which broke
down the Kantian antithesis between Sensibility and Understanding by showing or attempting to show the continuity of instinctive sensory reactions and the more guarded apprehensions of
reflective thinking.

The very different aim of the evolutionist from that of his opponents—
the latter seeking to account for the *resemblances* of mental actions in beings supposed to be radically different in their mental constitutions, while
the former seeks to account for the *differences* of manifestations in funda-
mentally similar mental constitutions—gives, in the theory of evolution, a
philosophical *rôle* to the word "instinct," and to its contrast with intelligence, much inferior to that which this contrast has had in the discussions
of the mental faculties of animals.[48]

In other words, Wright in his "psycho-zoology" did not
make sharp distinctions, as vitalists and Kantian idealists did, between organic and inorganic, vegetable and animal forms and
functions, voluntary and involuntary, instinctive and intelligent

motives and actions; but regarded these as rough divisions in a continuous series of evolutionary processes: "*Habits* properly so called, and dispositions, which are the inherited effects of habits, are not different in their practical character or modes of action from true instincts." [49]

That our highest flights of knowledge are the evolutionary outcome of our most inborn or instinctive predispositions—of feeding (for the physical sciences) and of breeding (for the psychical sciences)—is a characteristic feature of Peirce's theory of the genesis of knowledge. But Peirce's approach to this theory, as we have seen above, was that of a logician primarily:

> The highest kind of symbol is one which signifies a growth, or self-development, of thought, and it is of that alone that a moving representation is possible; and accordingly, the central problem of logic is to say whether one given thought is truly, *i.e.*, is adapted to be, a development of a given other or not.[50]

For both Wright and Peirce, the use of signs and symbols constitutes thought and explains the continuity of man's intellectual evolution.

Two more citations illustrating the profound influence of evolutionism on Peirce's logic will show how the relativity and temporality of knowledge invaded even the elementary distinctions of formal logic. In discussing the denotation and connotation of terms, Peirce calls attention to the fact that the distinction between the logical *breadth* and *depth* of concepts—as he preferred to designate it—was not an absolute one intrinsic to a dyadic relation between signs and objects, but that both denotation and connotation take their origin in the triadic relation between a sign, its object, and its interpretation; "and furthermore, the distinction appears as a dichotomy owing to the limitation of the field of thought, which forgets that concepts grow, and that there is thus a third respect in which they may differ, depending on the state of knowledge, or amount of information." [51]

The second citation is taken from the branch of logic which Peirce, after Boole and De Morgan, did most to develop, namely, the application of mathematical logic to the calculus of probabilities.

By means of this simple calculus [of classes], he [Boole] took some great steps towards the elucidation of probable reasoning; and had it not been that, in his pre-Darwinian day, the notion that certain subjects were profoundly mysterious, so that it was hopeless, if not impious, to seek to penetrate them, was still prevalent in Great Britain, his instrument and his intellectual force were adequate to carrying him further than he actually went.[52]

Peirce is here chiding Boole for failing to go as far as Peirce went when he maintained: "To say, therefore, that thought cannot happen in an instant, but requires a time, is but another way of saying that every thought must be interpreted in another, or that all thought is in signs";[53] or as Peirce says in another place: "The woof and warp of all thought and all research is symbols, and the life of thought and science is the life inherent in symbols; so that it is wrong to say that a good language is *important* to good thought, merely; for it is the essence of it." [54]

John Dewey, to whom I am grateful for indicating the significance of these last two passages for Peirce's theory of logic, is faithful to the evolutionist background of Peirce's later (post-Darwinian) logical writings when he interprets Peirce's theory of signs to mean

that in the course of cosmic or natural evolution, linguistic behavior *super*venes on other more immediate and, so to say, physiological modes of behavior, and that in supervening it also *inter*venes in the course of the latter, so that through this mediation regularity, continuity, generality become properties of the course of events, so that they are raised to the plane of reasonableness. For "the complete object of a symbol, that is to say, its meaning, is of the nature of a law." [55]

The life of reason is the evolution of ideas which are as real, in their symbolic functioning and influences on our desires and actions, as particular events. Ideas are the meanings of events, and evolve in time with the growth of knowledge.

Peirce's reference to the avoidance of metaphysical issues in Great Britain in the middle of the nineteenth century, and to the "pre-Darwinian" character of scientific thinking at the time of Boole's[56] works on mathematical logic, indicate that Peirce held the British nominalists responsible for holding back the clue to the metaphysical basis of logic supplied by evolutionism. In other words, in his evolutionism Peirce saw the chain of ideas with which to link his scholastic realism to the reorientation of the logic of science, misled for five centuries by Ockhamite individualism.

Today [1901], the idea uppermost in men's minds is Evolution. In their genuine nature, no two things could be more hostile than the idea of evolution and that individualism upon which Ockham erected his philosophy. But this hostility has not yet made itself obvious; so that the lion cub and the lamb still lie down together in one mind, until a certain one of them shall become more mature. Whatever in the philosophies of our day (as far as we need consider them [in relation to the idea of a law of nature]) is not Ockhamism is evolutionism of one kind or another; and every evolutionism must in its evolution eventually restore that rejected idea of law as a reasonableness energizing in the world (no matter through what mechanism of natural selection or otherwise) which belonged to the scholastic modifications of it by Aquinas and Scotus.[57]

Or again, Peirce, in order to defend the cosmic reality of purposiveness, remarks: "Almost everybody will now [1902] agree that the ultimate good lies in the evolutionary process in some way." [58] He then goes on to refer to his logical and metaphysical doctrine of continuity, "synechism," or the doctrine "that the coalescence, the becoming continuous, the becoming governed by laws, the becoming instinct with general ideas, are but phases of one and the same process of the growth of reasonableness. This is first shown to be true with mathematical exactitude in the field of logic, and is thence inferred to hold good metaphysically." [59] This evolutionary growth of reasonableness includes the "procedure" of pragmaticism "as a step."

Why should the growth of habits include the procedure of pragmaticism as a step? The growth of reasonableness, as explained above, simply means, apart from Peirce's cosmic panpsychism, the gradual predominance of systematic habits or conscious methods over random, trial and error efforts to solve the problems of that struggle among ideas which characterizes the history of human thought. That part of the history of human thought which deals with the analysis of the meanings of ideas and the discovery of what truth there is in them, belongs to the history of science. Peirce's pragmaticism is a method of making the meanings of our ideas clear by asking us to consider which of their *logical* consequences we are willing to act on or adopt as a possible mode of action. "Consider what effects that might conceivably have practical bearings we conceive the object of our conception to have. Then, our conception of these effects is the whole of our conception of the object." [60]

We cannot deduce conceivable effects without adopting leading principles as the relatively fixed characters of experience. And so we cannot grow in reasonableness unless we establish orderly habits of logical procedure in our conceptions. But intelligence consists in the habit of breaking old, inadequate habits and forming new, flexible ones in order to cope with the never-ending novelties of experience. No unthinking mechanism can be intelligent or grow in reasonableness.

The pragmaticism of Peirce clearly emphasizes the internal or logical function of those "doings" or operations which are said to be the "practical" method of clarifying the meanings of ideas. The essay "How To Make Our Ideas Clear," which James and Peirce regarded as the first statement of pragmatism (without that name appearing in the article), was originally an illustration of the logic of science generalized to serve as an aid to reflection in all perplexing *intellectual* matters, such as those of metaphysics (free-will, God, and immortality). The practical test of clarity is for Peirce essentially a conceptual and logical one: an idea is clear when we understand its *conceivable* effects or the logical consequences necessitated by adopting it as a premise or rule for the resolution of a problem. There is no question here of any individual's psychology, or of the immediately felt effects of pleasure or pain associated with the entertaining of ideas, or of brusque action* ensuing directly upon the acceptance of an idea. Thinking does not aim in science to provide directions to perform immediate actions or to justify fixed beliefs or to observe and enumerate successive sensations. "Thought is a thread of melody running through the succession of our sensations." [61] Thought is essentially relational, and the error of nominalism consists in de-

---

* "The doctrine [of James] appears to assume that the end of man is action —a stoical maxim, which to the present writer at the age of sixty, does not commend itself so forcibly as it did at thirty" (*C.P.*, 5.3, written in 1902).

nying the objective reality of relations. The logical ordering of relations is objective in the sense that relations are not figments of the imagination but lines of the activity of thought converging upon truth. For all those engaged in scientific or scholarly research, the meanings of truth and reality are explicitly defined by Peirce in a now classical sentence: "The opinion which is fated to be ultimately agreed to by all who investigate, is what we mean by the truth, and the object represented in this opinion is the real." [62] This definition of "truth" does make it depend on what *would* be everybody's opinion in the long run if everybody were to adopt the pragmaticist's method of discussing, investigating and clarifying the ideas of truth and reality. One of the consequences of pragmaticism is that it may take an inconceivably long time for the conceivable effects of an idea or thing to become realized. In the phrase "in the long run" there lies a certain vagueness that made it easy for Peirce to assimilate in his own last writings some of the mystical elements of religious idealism in the metaphysics of Royce and Henry James the elder, both of whom Peirce admired and whose influence he admitted. [63]

Peirce accepted Kant's synthetic a priori forms (causality, relation, and so on, and other universal and necessary types of assertion) as *regulative* principles of thought, but not as constitutive of *phenomena* merely. There is in Peirce a Schellingian tendency to regard the forms of thought as constituting the forms of reality, and therefore whatever is regulative of thought is bound to be transformative of things or events, for the latter embody feelings or ideas. A second difference between Peirce's and Kant's metaphysics resides in the temporal and evolutionary character of thought and reality for Peirce, Kant's forms being eternal and fixed in the immutable constitution of the human mind. Regulative principles, or as Peirce called them, "leading principles," are, for him, of the nature of habits. In his cosmological use of this term, Peirce commits himself to the organismic view that nature—including that part of it we call human mind —grows like a living creature, forming new habits, modifying or breaking old ones, and showing her maturity and increasing reasonableness by abandoning gradually, as time goes on, the unpredictable waywardness of her wild tychistic youth and acquiring more stable continuity and self-control in the form of self-imposed laws. This sort of metaphysical evolutionism is a far cry from Darwin's evolutionary hypothesis, in which the physical environment disciplines by threats of extermination any variations from existing species which are not adaptable to external conditions.

Against the mechanistic view of Spencer's theory of evolution, which attempted to guarantee human progress, Peirce argued that it is impossible for progress to come about by mechanical necessity. The Hegelians, with their dialectical form of necessity, were also wrong in denying progress through individual effort. "A pseudo-evolutionism which enthrones mechanical law above the principle of growth is at once scientifically unsatisfactory, as giving no possible hint as to how the universe has come about, and hostile to all hopes of personal relations to God." [64] Peirce chose a daring way to cut the Gordian knot of the controversy between mechanistic and vitalistic types of evolutionism. He denied both the absoluteness of natural law and the infallibility of human knowledge. The absolutist made natural law itself incapable of growth and purpose, while Peirce argues that if laws, being absolute, cannot grow, they have no reason for being, other than their sheer eternal presence or inexplicable creation *ex nihilo*. The more rational view is to regard law as continuous with the natural forms of existence which are always growing and working out definite ends, not necessarily those of any mind. The only difference between human ends and other ends in nature is that we have more control over the former:

> Now who will deliberately say that our knowledge of these laws [of gravitation, elasticity, electricity, and chemistry] is sufficient to make us reasonably confident that they are absolutely eternal and immutable, and that they escape the great law of evolution? Each hereditary character is a law, but it is subject to development and to decay. Each habit of an individual is a law; but these laws are modified so easily by the operation of self-control, that it is one of the most potent of facts that ideals and thought generally have a very great influence on human conduct. That truth and justice are great powers in the world is no figure of speech, but a plain fact to which theories must accommodate themselves.[65]

Peirce means by law any constancy or regularity, and since the latter means persistency or reality, the evolution of the real world implies the evolution of laws as well as of species of organisms. Since thought is adapted to reality in the sense in which the "glassy essence" of mind provides a mirror for being, "the end of being and highest reality is the living impersonation of the idea that evolution generates." [66]

If God is the end of being and highest reality, then God is a growing creature of evolution. James, Bergson, S. Alexander, and H. G. Wells were thus anticipated not only by Schelling,[67] but also by the self-styled Schellingian, Peirce, in his evolutionary religion of the 1890's. As for Wright and James, religion for Peirce was valuable as a practical support of moral ideas. "In its higher stages, evolution takes place more and more largely through self-control . . ." [68] Since self-control lies behind logical reflection, and operates at the center of all ethical decisions, evolution tends toward an ethical goal.

Thus there are many *obiter dicta* in Peirce's accounts of his own philosophical procedure as he evaluated it which make us doubt the pure logicality of his evolutionism. We have seen how Peirce went so far as to maintain that "in biology, that tremendous upheaval caused in 1860 by Darwin's theory of fortuitous variations was but the consequence of a theorem in probabilities." [69] But six months later, in a letter,[70] Peirce reveals an entirely different mainspring of his evolutionism, which is far from being anything like a theorem of probabilities; it not only violates Wright's conception of the impersonal neutrality of science, but also suffers from the inconsistencies indicated in Professor Lovejoy's comments[71] on the following passage from Peirce's letter:

> To me there is an additional argument in favor of this theory of objective chance—I say to me because the argument supposes the reality of God, the Absolute, which I think the majority of intellectual men do not very confidently believe. It is that the universe of Nature seems much grander and more worthy of its creator, when it is conceived of, not as completed at the outset, but as such that from the merest chaos with nothing rational in it, it grows by an inevitable tendency more and more rational. It satisfies my religious instinct far better; and I have faith in the religious instinct.

James agreed that the creativity of nature should not be conceived as completed, but his empirical self was not satisfied with Peirce's cosmic conception of an inevitable tendency of chance to evolve toward a remote rational goal. Instead, James, like Wright, looked for the springs of religious faith in man's emotional nature which, in seeking self-expression vests an otherwise neutral evolutionary process with human significance.

CHAPTER FIVE

Darwinism in James's Psychology and Pragmatism

1. JAMES'S EARLY INTEREST IN DARWINISM

Against the opposition of the spokesmen of orthodox religion, scientists developed the evolutionary conception of nature as continuously producing its own myriad and changing variety of forms. In the sciences of man evolution was to encounter continued resistance from the older and more powerfully entrenched traditions that dominated mental and moral philosophy. Evolutionary studies of man first found it necessary to combat antievolutionary theologies and metaphysics not only in nonscientists but also in scientists reared on those traditions. Then we find after 1870 ingenious forms of a *Pax Christiana* contrived by sincere and learned American thinkers at Harvard like Charles Peirce, F. E. Abbot, and John Fiske. Their metaphysical ingenuity was spent in elaboration of scientific theisms, although they did keep the light of reason burning hopefully in the Cimmerian darkness of their evolutionary theologies. The other members of the Metaphysical Club turned to the more substantial scientific studies of human evolution. William James was fascinated by the evolution of the brain and nervous system and by Darwin's account of the physiological expression of emotions in man and animals, along with other Darwinian views of the origin of human traits. Early in his teaching career at Harvard he used Spencer's texts on the synthesis of physical and social evolution. In a letter to President Eliot, written in 1875—a few months after Wright's death— James urged the need for a new type of chair in psychology, one that should be occupied by a philosophical scientist, like Wundt or Lotze, trained in nervous physiology and informed of the recent "archaeological" studies in the development of language and

social institutions.° These anthropological studies were still combating the influence of theologians and metaphysicians who upheld fixed, eternal, and "higher" truths—higher, that is, than the
tentative piecemeal conclusions of less pretentious scientific investigators. These silent, slowly advancing conquerors of truth
did not reassure those who clung to a medieval faith, or to an
eighteenth-century faith in the natural harmony of social and
selfish interests, guaranteed by laws of social progress. The val-
iant ambition of the founders of pragmatism was to restore man's
faith in himself, after science had exposed the mythical character
of theologies and of metaphysical systems that based faith and
morals on "self-evident" and "eternal" laws.

One has only to compare the different ways in which William James and his father grappled with the problems of the
mind in order to sense the profound change brought about within
a single generation by the scientific approach to man after Darwin's work on the *Origin of Species*. When Henry James, Sr.,
about 1840, wrote to his friend, Joseph Henry, the Faraday of
America, for some scientific evidence of eternal moral and religious truths of the universe, he received in reply no satisfactory
assurance that physical science was ready to uphold the heaven
and hell of Swedenborg. With a passion equal to that of Henry
James, Sr., and his generation, but with greater scientific skepticism, William James and his friends in the Metaphysical Club
debated whether the new biological discoveries of Darwin could
really do for the human world what the older metaphysical doctrines had promised but failed to do. It is not surprising that as
a result of his Darwinian studies in physiology and psychology,
William James did not emerge with any final answers to the old

---

° "A real science of man is now being built up out of the theory of evolution and the facts of archeology, the nervous system and the senses. It has already
a vast material extent, the papers and magazines are full of essays and articles having more or less to do with it." Letter of James to President Eliot, December 2,
1875, in Perry, *Thought and Character of William James* (Boston: Little, Brown,
1935), II, 10–11. James Mark Baldwin observed that "early American psychology
was written by theologians or educators or both in the same person"; quoted by
Jay Wharton Fay, *American Psychology Before William James* (New Brunswick,
N. J., 1939), preface. When Henry Holt asked John Fiske in 1878 to write a book
on psychology, Fiske recommended William James who "knows more about it
than I do. He has been studying little else for several years." *Letters of John Fiske,*
edited by his daughter Ethel F. Fisk (New York, 1940), p. 371.

metaphysical problems. But he did revolutionize American psychology and philosophy by carrying evolution as far as he could in both fields, and by clearing the ground for a candid recognition of our scientific ignorance of the depths of human nature and of the world from which it emerged and is still emerging.

William James in his own empirical way did what Kant tried to do in an a priori way, namely, to show the limits of science in order to make room for faith—not an unintelligible metaphysical faith in an unknowable beyond experience, but a clear-eyed pragmatic faith in the individual as the chief repository of the hopes of humanity. James's pragmatism, though essentially resting on a belief in the genuine efficacy of human effort, did not appeal to a Schopenhauerian annihilation of desire or to a blind Nietzschean will to power. On the contrary, James argued for a more liberal and humane conception of the mind's dynamic power to ameliorate man's sorry lot. That neither physical nor biological science sealed man's fate or destined him to passive resignation in a closed universe was one of the chief moral and metaphysical conclusions of James's great psychological work. We shall notice in some detail how he drew on Darwin in order to obtain scientific support for his moral thesis. Without going too far beyond his *Principles of Psychology,* we shall discern the evolutionary germs of those ideas which he later sought to defend metaphysically: a functional view of the mind, temporalism and pluralism in the emotional, volitional, and intellectual life of the mind, and a moral and metaphysical individualism.

The relation of Darwinism to the development of James's philosophy will not be found set forth in any one piece of James's psychological or philosophical writing. As Professor Ralph B. Perry remarks in his definitive work on James, "the influence of Darwin was both early and profound, and its effects crop up in diverse and unexpected quarters." [1] Thanks to Professor Perry's magnificent and thorough portraiture of the thought and character of the most influential and universally beloved of all the founders of pragmatism, my task of bringing together the effects of Darwin's idea of evolution on James has been greatly facilitated. With Professor Perry we must discriminate an early positivistic phase of James's idea of evolution. In this phase James pitted himself against his anti-Darwinian teacher of zoology, the famous Louis Agassiz. Next, under the influence of Chauncey Wright and Charles Peirce, James began to criticize Spencer's mechanical evolutionism, but still clung to Darwin's theory of spontaneous variations despite Peirce's metaphysical criticisms. Finally, we have the metaphysical phase of James's later years, when he espoused Peirce's radical tychism: the extension of spontaneous chance to the whole of nature and its laws. It is easier to make out the positivistic and anti-Spencerian Darwinism of James's *Psychology* than the metaphysical phase, for the latter was never fully worked out. The metaphysics of evolution unfortunately did become entangled with James's earlier interest in morals and method. For we shall find James arguing with Chauncey Wright in 1875 about the duty or right to believe, long before he attempted a pluralistic and tychistic metaphysics which would provide room for it. Also we shall find James defending on Darwinian grounds the hereditary character of moral and intellectual predispositions long before he attempted a metaphysic of "radical empiricism" to get at the ultimate constituents of the world. Finally, the pragmatic method of clarifying the *meanings* of controversial terms and theories antedates his metaphysical flirtations with panpsychism, anti-intellectualism, and "crass supernaturalism," ideas with which James played but never developed fully.

Our main purpose now is to trace James's use of the Darwinian idea of evolution in his magnum opus, *The Principles of Psychology*, in order to discern its relations to the development of his pragmatism. In the background are the passionate exchanges of argument that took place between him and his youthful intellectual companions, Chauncey Wright, Charles Peirce, Oliver Wendell Holmes, Jr., and John Fiske, over the philosophical significance of evolution. We must remember that James did not give the doctrine of pragmatism its first public mintage and currency until 1898. Though he credited Peirce with having formulated and baptized the doctrine in the early seventies, we know that Wright, Holmes, and Peirce did not accept James's version of the doctrine. They early and severely criticized James's doctrine of the will to believe for putting man too close to the center of the evolving universe. That is why among the founders of pragmatism there were varieties of *application* of the method of determining the *meanings* of ideas by examining their temporal consequences or evolutionary effects on thought and conduct.

James adhered closely to the observational evidence supporting Darwin's hypothesis, as Chauncey Wright did, for both were critical of the dogmatic claims of those who tended to make a gospel of evolutionism (Spencer, Fiske, Abbot, and Peirce). But the variety and complexity of nature which Wright called "cosmic weather" and Peirce "tychism," became for James the metaphysical ground of the theory of an open universe and individual moral freedom. What James as a metaphysician finally retained of evolution, namely, the ideas of temporalism and spontaneous variation, served him persistently in his defense of the primary importance of individual experience and personal freedom. That is the Ariadne's thread to James's philosophy of evolution. The elusive but genuine character of individual spontaneity in both the external world and in man is in James's view of evolution epitomized by "saltatory" mutations,[2] original, spontaneous, irreducible phases of experience. James in his metaphysics dramatized the external world of sensations and the inner world of rational, moral, aesthetic, and religious sentiments. Take these spontaneous variations and creative impulses as you find them, and you have the ingredients of James's faith in the sufficiency of immediate experience, despite its transiency, and in the will to believe, despite the chilling, paralyzing doubts of scientific skepticism. Scientific and philosophical ideas become merely abstractions, useful only as intellectual instruments or convenient fictions to aid the individual to find his way among the complex particulars that flow into and out of his stream of experiences. The arrogance of metaphysical evolutionism is due to its attempt to substitute scientific abstractions for the more deeply felt flux. James's critique of evolutionism brought him later to the center of the *fin de siècle* attack on scientism, in which fight he welcomed the eloquent intuitionism of Bergson. In combating evolutionism as a new form of dogmatism in which science replaced theology, James at first defended scientific method because he felt it had been subjected to abuse by dogmatic, speculative evolutionists like Spencer. Hence he sought both a general theory of the method of clarifying all generalizations and a criterion of truth that would do justice to the specific *differences* that are felt in scientific and ethico-religious experiences. Perceptions felt immediately to be so different must make a practical difference to our conduct in exploring the meanings of ideas. These meanings become clear when considered as guides to conduct. Ideas are in James's dynamic psychology incipient expressions of our active natures. Hence, unless we act on an idea, imaginatively or actually, we do not know what its meaning fully portends; and only if we are *free* to enact ideas can we perceive their meaning. Whence, the will to believe doctrine—a doctrine peculiar to James's pragmatism and expressive of his humanism—that we can create certain kinds of human facts if we go ahead and act on their desired coming into existence. The theory of truth implicated by James's will to believe aimed to humanize science and fortify individual morality against scientific skepticism and neutral indifference. The meaning of an idea grows out of the particular effects we perceive when we act on it; truth is what happens to ideas when they fit our experiences dynamically, that is, when the flux of experiences becomes adapted to variations produced by our individual efforts. This temporal growth and fitness or adaptability of experience to new ideas in man's precarious struggle with his environment is in keeping with the Darwinian theory of natural selection. We need only to combine it with James's firm belief in the active and spontaneous character of the individual in order to see how the ideas of evolution, tychism, and a will to believe doctrine came together in James's early pragmatism.

> To attain perfect clearness in our thoughts of an object . . . we need only consider what . . . sensations we are to expect from it, and what reactions we must prepare . . . The ultimate test for us of what a truth means is indeed the conduct it dictates or inspires. But it inspires that conduct because it first foretells some particular turn in our experience which shall call for just that conduct from us.[3]

There is in James's pragmatic test of truth an emphasis on sensations rather than bare abstractions, on *our* reactions to external objects rather than on passive reception, and on individual freedom rather than general environmental determinism. "Radical empiricism" was James's way of demanding that we look always to the inexhaustible variety and details of immediate experience for the meaning of reality *for us* as centers of activity in the flux. The *particular* turn which we can give to our experience of objects endows our sensations of them with new meaning. The artistic metaphysician in James clung to that psychological fact throughout all his philosophizing over evolution, but it produced an internal clash of ideas with his scientific respect for Darwin's general hypothesis that the individual is at the mercy of environmental forces. The tough-minded positivistic medical student and physiologist was early at war with the tender-minded moralist and individualist. It was the latter strain which predominated as the years went by after 1890 and took James farther away from his physiological studies and deeper into the thicket of metaphysical questions. The public seemed to be more concerned with what science, particularly evolutionary studies, implied concerning the old questions of God, freedom, and immortality, than with the logical validity of evolutionism. James as a humanitarian responded more eagerly to the larger vital public interest than to specialized preoccupations of "scientific" philosophers, with whom he grappled when their theories seemed to him to threaten the primacy of the individual's immediate experience and unquenchable thirst for freedom.

In studying the effect of Darwin's ideas on James's pragmatism, we must continually keep before us the two aspects of Darwinism—in other words, of the natural selection theory: (1) the conception of random and therefore of spontaneous "chance" variations; (2) the conception of the action of environmental conditions in selecting those variations having survival value and vigorously eliminating all others. It was characteristic of James (and, in one of his veins of thought, of Peirce) to make much of the first, with its possibly (though, in fact, somewhat equivocally) indeterminate implications. Most of the Darwinians laid the emphasis upon the second, in the theory of the formation of species; factors external to the individual variant were all important in determining whether it (or its offspring) would survive, or not. So also in the un-Darwinian Spencer: evolution is "the continuous adjustment of internal relations *to* external relations." James may perhaps be said to have insisted, at times, that there is also a process in which external relations are adjusted to *internal* relations —that is to say, to the spontaneous variations which differentiate the individual. There is thus not only a "loose," not predetermined, nonmechanical aspect in the evolutionary process, but this manifests itself as a causal factor in the process, is dynamic and efficacious. The "spontaneous" variations are not merely passive, raw material to be weeded out by external, environmental factors; they, so to say, once they have emerged, do some weeding on their own account, react *upon* and, at least in some slight degree, *create* their environment. In human affairs faith in a fact, the will to believe, often helps to create the fact, James boldly argued.

On the other hand, the instrumentalist strain in pragmatism seems akin to, and perhaps partly derivative from, the second aspect of Darwinism. For the strict instrumentalist, mental functions—ideation, perception, judgment, reasoning— are *merely* means of adjusting "internal" to "external" relations; they are nothing but aids—superior to teeth and claws—to survival. Though this way of thinking—which long antedates Darwin—probably had some influence on James, in the main he was antipathetic to it; there are for him originally chance variations which have no *positive* survival value, but nevertheless survive; there are even some which are in some degree (though not to a decisive degree) unfavorable to survival. And in man these, so to say, completely non-Darwinian values are for James the supreme values—the truly "valuable" things. There is in his thinking a considerable interplay and wavering between these various motifs, and others not here mentioned. We must remind the reader that by "Darwinism" James meant not "evolutionism," in other words, the metaphysical generalization of organic evolution, but the natural-selection hypothesis.

Evidence of the linkage between James's use of the two central Darwinian ideas of chance variations and adaptation and his earliest expressions of pragmatism will now be adduced from James's early writings, culminating in his first book, *The Principles of Psychology*. If we keep to our Ariadne's thread, namely James's profound love of original individuality and spontaneous freedom, we shall not go astray among his intellectual vacillations. This thread runs through all his dialectical controversies with his Cambridge friends (Wright, Peirce, Holmes, Fiske) and leads us through the thick growth of his pragmatism. The profusion of his ideas had been nurtured early in the stimulating milieu of a

gifted family of rich and varied interests,* ranging from Swedenborg and spiritual socialism through literary criticism and the
psychological novel to evolutionary physiology and psychical research. Though early trained as a medical student in physiological details, James never lost sight of the great moral problems.
He approached them, as his scientific interests gave way to his
deeper philosophical unrest, with a cosmopolitan and metaphysical sweep that elicited criticism and admiration from thinkers
of varied schools of thought, startled out of their dogmatic
slumbers by the freshness and vitality of his manner of seeking
the human significance of science, religion, and metaphysics. He
showed an unusual capacity for forming intellectual friendships
with nearly every leading thinker of his time. Starting with his
companions in the Metaphysical Club and other clubs in and
around Harvard, the list of those who were attracted to him by
personal acquaintance or by correspondence is a veritable guide
to late nineteenth-century American and European thought. Professor Perry's work devotes at least one chapter to each of
the following: Oliver Wendell Holmes, Jr., Chauncey Wright,
Charles Peirce, Thomas Hodgson, Charles Renouvier, Thomas
Davidson, George H. Howison, Josiah Royce, Hugo Münsterberg,
Carl Stumpf, F. C. S. Schiller, John Dewey, Charles A. Strong,
Benjamin Paul Blood, Henri Bergson, James Ward. There were
hundreds of others from whom James learned a plurality of per-
spectives on life's problems. There are thousands who as his pu-
pils or readers have felt and still feel the engaging force of his
mind and his personal integrity, and the reader can do no better
than go directly to Perry's work for the full life portrait of James.
What he will find here is a highly selective glimpse of how one
idea, that of Darwinian evolution, functioned as a huge ganglion
or nerve center in a complex and sensitive mind reaching out to
all the vitally human questions of his day and ours.

2. POSITIVISM AND METAPHYSICS IN *The Principles of Psychology*

A serious-minded student at Harvard in 1877 complained in
the columns of the *Harvard Crimson* about the prospectus of a
new course to be given in the Philosophy Department by Wil-

* See F. O. Matthiessen, *The James Family* (New York, 1947).

liam James, an assistant professor of physiological psychology:
"Darwinianism," the complaint read, "is to be treated metaphysically, that is to say, . . . precisely as Darwin and his followers
say it should not be treated!" [4] This student's indignation was
due perhaps to the way James had interspersed his physiological
teaching with criticisms (inspired in part by Peirce) of Spencer's
philosophical speculations about evolution. There was also a
growing hostility to traditional metaphysics encouraged by positivistic followers of Darwin at Cambridge like Jeffries Wyman,
John Fiske, Charles Eliot Norton, and the recently deceased
Chauncey Wright. Wright's *Philosophical Discussions,* edited by
James's friend, Charles Eliot Norton, was published in 1877 by
subscriptions from the admirers of his acute exposure of the metaphysical and theological foibles of speculative evolutionists.
Charles Peirce often, in retrospect, pointed a scornful finger at
the agnosticism and positivism of the 1870's. At this time we
find more or less antimetaphysical minds of the first order in
every field of investigation: in the natural sciences, Maxwell,
Tyndall, Clifford, Mendeleev, Pasteur, Lister, Claude Bernard,
Koch, Helmholtz, Hertz, Kirchhoff, Mach, Stallo, Gibbs, Darwin;
in the social sciences there was a more metaphysical use of scientific data made by Comte, Buckle, Taine, Mill, and Marx; in
French realistic or "scientific" novels, Flaubert, Zola, and the
de Goncourts were drawing for details on physiology, archaeology
and anthropology. Leslie Stephen defended Huxley's "agnosticism" and a psychological interpretation of religion. G. H. Lewes'
physiological researches and biographical history of philosophy appealed to James. He recommended to his brother Henry,
Sainte-Beuve's method of "living with" an author, and praised
"literary psychologists" except when they tried to be too scientific.

It is not surprising, therefore, that William James underscored the positivistic approach in his *Principles of Psychology,*
a compilation of brilliant articles which he had written during
1878 to 1889 for *Mind,* the idealistic *Journal of Speculative Philosophy,* and literary periodicals like *Scribner's* and the *Nation.*
His only claim to originality was that his work might aid in the
development of a "natural science of psychology." [5] But this pref-
atory claim was more of a concession to the positivism of the
times than a consistently followed program. This vacillation in

James between science and metaphysics was immediately pointed out by every one of the leading reviewers of James's work, and evaluated, naturally, by each reviewer in terms of his own metaphysical preconceptions. James Mark Baldwin declared that James's conception of the stream of thought running parallel to the stream of physiological innervation

should bring comfort to spiritualists and confusion to their enemies. And the comfort becomes positive satisfaction when one reads his final chapter on "Necessary Truths and the Effects of Experience." Here he argues trenchantly against the "experience hypothesis," finds raw experience also inadequate, and finally puts into his "pulse of Thought" a cargo of rational principles. It is to be hoped that Professor James will some day write us a "Metaphysics." [6]

Baldwin's reference to the spiritualists was vindicated in the review of Josiah Royce, who interpreted James's method as Kantian transcendentalism: "Know the most recent empirical formulation of the facts upon which you mean to reflect, and *then* reflect as profoundly, and, if necessary, as transcendentally as you can: that is the lesson of Kant's own method." [7] Royce concluded that James's work was of great value to ethical students. [8] James Sully, the British positivist, praised the book but was disturbed mildly in noting that James's instinct for careful observation of fact and rigorous verification of theory was accompanied by a theological mysticism:

The mysticism appears more plainly in the whole doctrine of effort and free-will which, as the author would probably be the first to allow, is as much an aesthetic or ethic postulate as a psychologic assertion. Yet this Lotze-like leaning to a theological view of things is on the whole kept in praiseworthy subjection to properly scientific aims, and a man must have the smallness of a pedant to object to the occasional intrusion of a hint of large vistas beyond. [9]

James Ward, a friend of William James and an acute British psychologist with theistic leanings, put his finger on a fundamental dualism in James: "I should apply to you the word of Goethe: 'Es sind zwei Menschen in diesen Brust,' *u. s. w.* I shall some day play off James the psychologist against James the metaphysician, moralist, and human." James's reply to this criticism of his *Psychology* indicates that the dualism between the physiological and psychological was inescapable and that he had grown impatient with the exaggerated claims of so-called scientific psychology:

> Yes, I am too unsystematic and loose! But in this case [of psychology] I permitted myself to remain so deliberately, on account of the strong aversion with which I am filled for the humbugging pretense of exactitude in the way of definition of terms and description of states in the psychological literature. What does a human being really learn from it all beyond what he knew already by the light of nature? . . . There can be no psychology worth the paper it is written on (*except* the science of the correlations of brain states with objects known) until something sound in epistemology is done. *Pray* go ahead and do it! [10]

The germs of James's Darwinism and his early moral and metaphysical interests are seen lurking in more than one chapter of his *Principles of Psychology*. We find them proliferating abundantly in the chapters of his great book dealing with the evolution of the functions of the brain, instinct, emotions, will, reasoning, and in the last ten pages of the final chapter on "necessary truths and the effects of experience."

Evolution of the Brain-Functions.[11] Throughout the *Principles* James tried desperately to reconcile the mechanical explanations of science with the felt reality of purpose, not as belonging to two worlds, one disclosed to sense experience and the other to a higher transcendental reason, but as inseparably joined. What interests us is that James invoked evolution as the process that did the joining. The brain and the nervous system he called "organs of consciousness" because consciousness which is always ruled by the ends of desire needed organs as means for achieving intelligently whatever would realize its ends. Feelings, preferences, and intelligence thus reside in the very action of all nervous centers: "Like all other organs . . . [the nervous centers] evolve from ancestor to descendant, and their evolution takes two directions, the lower centres passing downwards into more unhesitating automatism, and the higher ones upwards into larger intellectuality." [12]

The *modus operandi* of the evolution of the brain's functions is not discussed by James until the last chapter, in which he comes out against the Lamarckian theory of the inheritance of acquired characteristics and defends Darwin's view of the preservation of lucky variations. The reflex actions through the spinal cord become more and more automatic,

whilst on the contrary those functions which it benefits the animal to have adapted to delicate environing variations pass more and more to the hemispheres, whose anatomical structure and attendant consciousness grow more and more elaborate as zoological evolution proceeds . . . Some such shadowy view of the evolution of the centres, of the relation of consciousness to them, and of the hemispheres to the other lobes, is, it seems to me, that in which it is safest to indulge. If it has no other advantage, it at any rate makes us realize how enormous are the gaps in our knowledge, the moment we try to cover the facts by any one formula of a general kind.[13]

Knowledge of the *intimate* working of the brain must be relegated to the physiology of the future.[14] Out of the shadowy view of the evolution of the brain's functions there emerges the central psychological fact about man: he is a creature of *preferences, desires, action,* and *will.*

To explain the irritability of the nervous system,° James alludes[15] skeptically to Fiske's *Cosmic Philosophy* (1874) and Spencer's *Principles of Biology* (1860) and *Principles of Psychology* (1855). Their mechanical explanations of the nervous system as a mass of matter in different states of tension and paths of action subject to the effects of nutrition until equilibrium is attained—all these Spencerian notions impressed James as concealing under a great show of precision much that was vague, improbable, and even self-contradictory. The more clear-cut and patent facts about the nervous system open to observation are its plasticity and dynamic habit-forming tendencies. What resists our efforts at first may yield and become automatic by exercise, so that habits newly formed may liberate consciousness to attend to creative acts of greater complexity. Dr. Carpenter's *Mental Physiology* (1874) was perhaps the chief source of James's theories of habit and ideo-motor action.[16] That all thinking was ideo-motor in character is an essential part of what

° That our very nerve fibers in a dim way feel, prefer, and desire was also the view of George H. Lewes, in his *Physiology of the Common Life* (1860), and of other physiologists impressed by the adaptive power of even the simple reflexes of the frog.

James's pragmatism imported into philosophy; hence, Carpenter's work, from this standpoint, shares with those of Bain, Lewes, Lotze, and Fouillée the credit for nourishing the psychological roots of James's pragmatism.[17]

Instincts. James's definition of instinct continues to show his philosophical concern to break down the opposition between purposive and mechanical actions: *Instinct is action towards ends without foresight and without previous learning.* If we find processes of reaching for ends by reflex actions operating on an enormous scale in the animal world, then we should not hesitate to impute the same to humans. Paul A. Chadbourne's popular work on *Instinct* (1872), to which I have alluded in my introductory chapter, is the first cited by James in support of the notion that there is no organ without a native aptitude for its use by its possessor.[18] However, James criticizes Chadbourne's abstract naming of purposes like "self-preservation, fear of death, love of life, maternal love," for failing to respect "the strict physiological way of interpreting the facts. . . . *The actions we call instinctive all conform to the general reflex type* . . . The nervous system is to a great extent a preorganized bundle of such reactions . . ."[19] The preorganization is evolutional, and instinct is not, as Chadbourne's theological theory would have it, a form of a priori knowledge. James ridicules Chadbourne's view that the bird with a gland for the secretion of oil *knows* instinctively how to press the oil from the gland and apply it to the feather, or that the hawk *knows* by instinct how to wield his talons. Yet when Bergson's *L'Évolution Créatrice* appeared with the same sort of arguments, in 1905, James had forgotten his own reflex theory of instincts and espoused an intuitionistic theory of knowledge by "pure experience," in his *Essays in Radical Empiricism.*

Returning, however, to James's early Darwinian and reflex theory of purposive instinct, we find him then especially impressed by the physiological details of structural adaptations in nature: "Even so there are no bounds to the minuteness of adaptation in the way of *conduct* which the several inhabitants of each nook and cranny of creation display."[20]

How can we explain these marvelous and intricate adjustments of animals to the many demands of their environment? James's answer does not, like Chadbourne's, make reference to the beneficence of God. It is an avowedly anthropomorphic answer which goes to the root of James's personalistic philosophy, and is far from Darwinian: "We can only interpret the instincts of brutes by what we know of instincts in ourselves." [21]

This openly anthropomorphic defiance of mechanistic science obviously goes counter to the way in which Darwin postponed for a dozen years his application of natural selection to man, for *The Descent of Man* and its principle of sexual selection was not put forth until 1872; in *The Origin of Species* there was scarcely any explicit reference to the origin of the human species other than the bare statement "Light will be thrown on the origin of man and his history" (p. 488). But James's early view was an attack on metaphysical abstractions like materialism rather than on science. He likewise attacked idealistic theories of human nature that minimized the life of the impulses and made pure reason à la Kant and Hegel the essential characteristic of man. James claimed that both materialists and idealists erred in their views of instinct, the former regarding them as blind, the latter as invariable. His evidence is taken from Darwin's *Origin of Species*:

> In the instincts of mammals, and even of lower creatures, the uniformity and infallibility which, a generation ago, were considered as essential characters do not exist. The minuter study of recent years has found continuity, transition, variation, and mistake, wherever it has looked for them, and decided that what is called an instinct is usually only a tendency to act in a way of which the *average* is pretty constant, but which need not be mathematically true. Cf. on this point Darwin's Origin of Species.[22]

James's sense of the plurality and individuality of nature's ways appears in his Darwinian view of instincts: *"Nature implants contrary impulses to act on many classes of things,* and leaves it to slight alterations in the conditions of the *individual* case to decide which impulse shall carry the day."* [23] There are a great many more impulses in man than the rationalists have recognized. It is not reason which checks any impulse, but only a stronger impulse. Reason may excite the imagination so as to let loose the stronger, more or less advantageous impulse. There is no intrinsic value or supremacy in either instinct or reason, and no material antagonism between them.

There is a certain maturation or "transiency," as James called it, of the many instincts of man productive of his character which the educator should utilize:

> In all pedagogy the great thing is to strike the iron while hot, and to seize the wave of the pupil's interest in each successive subject before its ebb has come, so that knowledge may be got and a habit of skill acquired—a headway of interest, in short, secured, on which afterward the individual may float.[24]

The modifiability of the elementary impulses in time is an essential ingredient of James's temporalism. And another component of his later metaphysics, his pluralism, appears in the large number of instincts he found innate in man. Darwinism explained their biologic utility, and tychism their variety. *"No other mammal, not even the monkey, shows so large an array"* (James's italics).[25]

The long list of special human instincts (over thirty in number) arrayed by James reveals his desire to do justice to the specific, original, variable, and dynamic roots of human nature. These had been described in niggardly fashion by metaphysical psychologists who attempted to reduce the mind of man to a few general invariable faculties or traits endowed by God. The first half of the list contains James's observations on his own child's behavior; they are similar to Darwin's empirical reports of his children's psychogenesis.

Beginning with the reflex action of sucking by infants ("seeking the breast is acquired") and going on to the social predispositions of imitation, emulation, or rivalry, James's inventory of instincts shows little concern for their early modification by cultural influences. On the contrary, his whole theory of the native endowment of man is a form of cultural primitivism. Perhaps this is what Dewey means when he says that James was lacking in historical sense. For example, from the instincts of *pugnacity, anger, resentment,* James tries to derive and explain the tragedies of war, ignoring completely economic and political factors:

> *Pugnacity; anger; resentment.* In many respects man is the most ruthlessly ferocious of beasts. As with all gregarious animals "two souls," as

Faust says, "dwell within his breast," the one of sociability and helpfulness, the other of jealousy and antagonism to his mates. Though in a general way he cannot live without them, yet, as regards certain individuals, it often falls out that he cannot live with them either. Constrained to be a member of a tribe, he still has a right to decide, as far as in him lies, of which other members the tribe shall consist. Killing off a few obnoxious ones may often better the chances of those that remain. And killing off a neighboring tribe from whom no good thing comes, but only competition, may materially better the lot of the whole tribe. Hence the gory cradle, the *bellum omnium contra omnes,* in which our race was reared; hence the fickleness of human ties, the ease with which the foe of yesterday becomes the ally of to-day, the friend of to-day the enemy of to-morrow; hence the fact that we, the lineal representatives of the successful enactors of one scene of slaughter after another, must, whatever more pacific virtues we may also possess, still carry about with us, ready at any moment to burst into flame, the smouldering and sinister traits of character by means of which they lived through so many massacres, harming others, but themselves unharmed.[26]

In "The Moral Equivalent of War," published in 1910, James does not depart from the assumption that bellicosity is engrained in man, but adds the "solution": sublimate it in sports and international work projects. There is still no inkling of the powerful economic and political causes of international war, though there is no lack in James of deep moral indignation at their effects in degrading man. What James did contribute to the problem was the important psychological fact that human nature is plastic enough to modify its combative tendencies and call into play more coöperative impulses. A few more examples of James's account of the native predispositions of man will suffice to bring out the individualistic character of his Darwinism, due to his slighting the socio-cultural factors that transform the raw, plastic biological endowment of man.

When James says *the hunting instinct* has an equally remote origin in the evolution of the race," he does not thereby explain the large variety of objects or social forms of all sorts historically associated with hunting. He does mention anthropological data in connection with the instinct of *play,* but fails to explain why its diverse cultural manifestations should be determined by such a general instinct: "The lowest savages have their dances, more or less formally conducted. The various religions have their solemn rites and exercises, and civic and military power symbolize their grandeur by processions and celebrations of diverse sorts." [27] Play enters into these complex cultural forms as it does into "higher esthetic feelings" but *how* it enters is not explained by calling it an instinct, as James does. The scientific study of cultural forms is not furthered by James's view that they arise out of instincts which are ultimate, inexplicable elements of human nature. Sociability and shyness are instincts also, says James, referring to Galton's observations on the gregariousness of South African cattle and to Darwin's remarks on stage fright.[28] But cattle do not talk and stage fright is more than shyness. Again, the social complication of instincts is ignored by James's well-intentioned evolutionary moralizing on the instinct of *Love:* "As Darwin has shown in his book on the *Descent of Man and Sexual Selection* it has played a vital part in the amelioration of all higher animal types, and is to a vast degree responsible for whatever degree of chastity the human race may show." [29]

Whenever physiology does not suffice to distinguish one instinct from another, James will resort to Darwinian "variation." Thus the instinct of *sympathy* is for James—the most sympathetic of philosophers—a peculiarly human trait, not to be confounded with or derived from animal gregariousness: "*Sympathy* does not necessarily follow from the mere fact of gregariousness . . . and may be due to a random variation selected, quite as probably as gregariousness and maternal love are, even in Spencer's opinion, due to such variations." [30] Similarly, science and philosophy as activities are not based on the more practical instinct of curiosity about the objects of the immediate environment, an instinct "already pretty low down among vertebrates." The stimuli to scientific and philosophic thinking "are not objects, but ways of conceiving objects; and the emotions they give rise to are to be classed, with many other aesthetic manifestations, sensitive and motor, as *incidental* features of our mental life." [31] By *incidental*, of course, James means that art, science, and philosophy are not needed for the immediate practical end of biological survival.

It would be misleading to allow my accumulation of selected portions of James's work to give the reader the one-sided impression that James approached his study of man only from the Darwinian viewpoint. James never approached any subject or problem from a single point of view. I shall quote only one instance here, in closing my discussion of his account of instinct, though many more passages could be cited in which James dropped the whole evolutionary approach in order to go directly to experience. Referring to *constructiveness,* he says: "Of course the utilitarian origin of this instinct is obvious. But to stick to bare facts at present and not to trace origins, we must admit that this instinct now exists, and probably always has existed, since man was man." [32] Peirce was exasperated by James's habit of dropping all theories and referring to what is given. He deplored the lack of logical rigor in James's piecemeal functional analysis of human traits along Darwinian lines.

The last ten pages of James's opus are devoted to the defense of Darwin's as against Lamarck's theory. The latter attributed the origin of instincts to acquired and transmitted ancestral habits. James's concluding pages summarize the arguments and evidences on both sides of the questions concerning psychogenesis or the evolutionary explanation of the native endowment of man. He brought to the support of his nativism Weismann's *Essays on Heredity,* published in English in 1889.[33] Though most psychologists were still Lamarckian in their view of instincts as products of ancestral experience, James stuck to his profound conviction of the uneliminable originality and native peculiarity of each individual's constitution independent of ancestral acquisitions. Such originality and independence fitted well into James's moral doctrine of the free will to believe and his correlated metaphysics of an open universe. James preferred Darwin's theory of the natural selection of accidentally produced variations to the Lamarckian theory, because the latter meant the visiting of the virtues and sins of the fathers upon their offspring, whereas the Darwinian view gave each individual a greater latitude to vary from his parents.

The greatest flaw in James's Darwinian and physiological theory of man's original nature is an almost complete disregard of the social conditioning of "instinctive" behavior, a flaw which runs through his theory of emotions, will, reasoning, and the springs of historical progress. His moral individualism limited the scientific but broadened the metaphysical range of his vision.

Emotions. As early as 1867, James, while a medical student in Berlin, had noted in the margin of his copy of Lotze's *Medi-*

cinische Psychologie that emotions are "bodily reverberations."
There were many more anticipations of the James-Lange physiological theory of the emotions than James was aware of,[34] but
the older medical and philosophical descriptions of emotions,
even when physiological, were still couched in general terms and
abstract classifications. Guided by the accurately detailed descriptions of such emotions as fear, provided by Darwin in his
Expression of the Emotions in Man and Animals (1872), James
defended the view that all emotions are organic disturbances,
"rigidity of this muscle, relaxation of that, constriction of arteries
here, dilation there, breathing of this sort or that, pulse slowing
or quickening, this gland secreting and that one dry, etc." [35] It
was precisely because the older literary, philosophical, and medical writers were rarely attentive to the minute and recondite
physiological causes of emotion that they failed to describe adequately the rich gamut of its subtle overtones. The internal
blendings and fugues of emotion are as varied and interpene-
trating as the innumerable pulsations and reverberations within
our organisms. Pre-Jamesian psychology erred in its accounts of
the emotions as pre-Darwinian biology had in its rigid classification of species:

> The trouble with the emotions in psychology is that they are re-
> garded too much as absolutely individual things. So long as they are set
> down as so many eternal and sacred psychic entities, like the old immutable
> species in natural history, so long all that *can* be done with them is rever-
> ently to catalogue their separate characters, points, and effects. But if we
> regard them as products of more general causes (as "species" are now
> regarded as products of heredity and variation), the mere distinguishing
> and cataloguing becomes of subsidiary importance . . . Now the general
> causes of the emotions are indubitably physiological.[36]

Just as positivists argued that a physical object is the
totality of its sensible effects, so James has shown that an emotion is the totality of its organic manifestations.

> Can one fancy the state of rage and picture no ebullition in the chest,
> no flushing of the face, no dilation of the nostrils, no clenching of the teeth,
> no impulse to vigorous action, but in their stead limp muscles, calm breath-
> ing, and a placid face? The present writer, for one, certainly cannot . . .
> A purely disembodied human emotion is a nonentity.[37]

The physiological theory defended by Darwin and James is not metaphysical materialism or behaviorism. James's functional analysis of emotion is neutral, in Chauncey Wright's sense, with respect to introspective knowledge and valuation of emotional reactions. The organic nervous disturbances which cause or condition emotions do not alter the fact that they are inwardly *felt,* and judged instantly as good or evil. James tried hard from the start of his *Psychology* to stick to a positivistic program which would make psychology a natural science. That meant making no commitments concerning the ultimate nature of the mind, but it is scarcely true to say that the only generalizations he permitted himself to use were those drawn from observable physiological functions interpreted by Darwinian principles.

Some emotional expressions are accounted for by the principle of natural selection *"as weakened repetitions of movements which formerly* (when they were stronger) *were of utility to the subject."* [38] This evolutionary principle of utility was applied by Darwin in explaining the one-sided uncovering of the upper teeth in the snarl or sneer. "It is accounted for by Darwin as a survival from the time when our ancestors had large canines, and unfleshed them (as dogs now do) for attack." [39] Does the pupil of the eye become dilated or contracted in fear? James has doubts about a contemporary physiologist's hypothesis on this detail: "I must repeat, with Darwin, that we need more minute observations on this subject." [40]

James in one place out-Darwins Darwin by criticizing the latter's principle of "antithesis"—opposite movements for contrary feelings—as too speculative to serve as a *causal* principle. He did not hesitate to abandon the teleological Lamarckian tendencies which crop up in Darwin, and came out vigorously for a purely accidental element having no *inherent* utility in the life of the nervous system, so far as the origin of its enormous variety of reactions goes. Note the inclusion of the aesthetic life among quasi-accidental variations: "Sea-sickness, the love of music, of the various intoxicants, nay the entire aesthetic life of man we shall have to trace to this accidental origin. It would be foolish to suppose that none of the reactions called emotional could have arisen in this *quasi*-accidental way." [41]

James's Darwinian theory of the emotions clearly reveals

the early positivistic phase of his pragmatic philosophy. It marks his early refusal to accept any metaphysical theory in which disembodied feelings or mental faculties are divorced from the emotional undercurrents of the "stream of thought." He also rejected as too metaphysical the panpsychist theory of an evolutionary mind-stuff that made feelings the primordial dust from which our consciousness emerged. So long as James stuck to the tough-minded physiological view of emotions as bodily reverberations, he was able to steer clear of unverifiable speculations about the *ultimate Urstoff* initially present in the cosmic dawn of consciousness. It was in a subsequent more tender-minded metaphysical mood that he flirted with evolutionary panpsychism,[42] but even then his early scientific scruples and individualism prevented him from espousing such an all-embracing metaphysical scheme. "Metaphysics," says James in his critical discussion of the mind-stuff theory—one of the most metaphysical chapters of the *Psychology*—"means nothing but an unusually obstinate effort to think clearly." [43] Does evolutionary psychology demand a mind-dust? The metaphysical thesis that it does so turns out to rest on contradictory premises: (1) that all mental states are compounded of *discrete*, atomic, primordial feelings, (2) that evolutionary *continuity* requires that consciousness in some shape must have been present in the very origin of things. If one has to make a choice between these two assumptions, James is certain that the first is false because of its implied mental chemistry of atomic mental states. Our mental states are experienced as "wholes" and are never truly resolvable into discrete things, psychical or material.[44]

James would rather leave open the metaphysical possibility of a soul that mysteriously interacts with the body than accept the unintelligible mind-stuff theory as a necessary supposition for evolutionary psychology. Chauncey Wright had shown how weak and attenuated is the metaphysical evolutionist's attempt to read Darwin's principle of natural selection and continuity of organic development into the remote astronomic and inorganic past of the universe. In Wright's "psycho-zoology"—title of an unborn treatise foreshadowed in "The Evolution of Self-Consciousness" (1873)—the natural science of psychology would employ Darwinian principles of natural selection on the verifiable phenomena of our mental states correlated with their physiological functions. So too, James restricted psychology as a natural science to giving an account of "particular finite streams of thought, co-existing and succeeding in time." Evolution in psychology would simply mean giving an empirically verifiable account of the *genesis* of mental states, and showing how they are useful in serving the interests of the human organism, without attempting to account for the ultimate origination of any of its spontaneous interests. Metaphysical theories of atomic sensations or a mind-stuff or "unconscious mental states" or an all inclusive Absolute Mind, are worthless as premises for a scientific psychology. By keeping to the functional interrelations of introspective and physiological data,

> our psychology will remain positivistic and non-metaphysical . . . the spiritualistic reader may nevertheless believe in the soul if he will; while the positivistic one who wishes to give a tinge of mystery to the expression of his positivism can continue to say that nature in her unfathomable designs has mixed us of clay and flame, of brain and mind, that the two things hang indubitably together and determine each other's being, but why or how, no mortal may ever know.[45]

Peirce's cosmic evolutionism which postulated a primordial chaos of unpersonalized feelings was regarded as a possible metaphysical theory by James, but out of place in an empirical psychology. The individualism of James would not sacrifice the integral character of present states of feeling or thought for any metaphysical theory, either while he wrote the *Principles* or when he departed from them in his later, more metaphysical phase. The temporalism of his pragmatism was rooted in the stream of thought, and his pluralism shows up in his conception of the various kinds of selves we all possess. Though natural selection leads us to desire the preservation of the material and social self, chance luckily also permits us spontaneously to strive for spiritual ends not directly related to the body or to society.

3. WILL, REASON, AND THE WILL TO BELIEVE

James's chapter on the will was originally published as an article, "The Feeling of Effort," in 1880, and as this title suggests, James attacked the classical and Kantian separation of will and reason from emotion. In line with his Darwinian theory of the emotions, James identified the feeling of effort with the kinesthetic feelings that invariably accompany so-called manifestations of will. Even the positivistic physiologists whom James admired, Bain, Helmholtz, and Mach, had not gone so far, but had with Wundt regarded will as a *prior* innervation of motor centers. James employed again a Darwinian notion of *selection* exercised by the efferent nerves on incoming sensations, a selection that would be guided spontaneously by our primary reflex, instinctive, emotional movements or interests. *Willing* and *doing* are one and the same psychical movement arising unaccountably from our spontaneous and deeply rooted ideo-motor make-up.

> We explain the mystery [of will and its effects] *tant bien que mal* by our evolutionary theories, saying that lucky variations and heredity have gradually brought it about that this particular pair of terms should have grown into a uniform sequence. Meanwhile why any state of consciousness *at all* should precede a movement, we know not—the two things seem so essentially discontinuous.[46]

There is a constant oscillation in James's mind between his respect for a scientific theory like Darwin's and his vivid sense of the concrete particularity of those aspects of immediate experience which elude scientific formulation, between the general laws of physiological behavior and the particular feelings disclosed to introspection. Profoundly aware of the need of correlating the two, James went to the roots of scientific theory and emerged with a nominalistic view that made the individual's sense-experiences and native modes of reaction the *fons et origo* of all theory, scientific, moral, or metaphysical. This nominalism distinguishes James's pragmatism from Peirce's. It reveals itself in a homely but striking passage in the chapter on "The Will" where James analyzes the reasoning of a drunkard struggling with himself to form a "reasonable" or "right conception" of his addiction. For James, as for Hume, the "right" conception in moral deliberation is one in harmony with our impulsive nature, but James added the Darwinian notion that in order to make the adjustment of a moral theory with powerful emotions there is an internal struggle of ideas attached to their names. The right idea survives only by crowding out all other names or words that obstruct the effort to face the meaning of "being a drunkard."

Where, however, the right conception is an anti-impulsive one, the whole intellectual ingenuity of the man usually goes to work to crowd it out of sight, by the help of which the disposition of the moment may sound sanctified, and sloth or passion may reign unchecked. How many excuses does the drunkard find when each new temptation comes! It is a new brand of liquor which the interests of intellectual culture in such matters obliges him to test; moreover, it is poured out and it is sin to waste it; or others are drinking and it would be churlishness to refuse; or it is but to enable him to sleep, or just to get through this job of work; or it isn't drinking, it is because he feels so cold; or it is Christmas day; or it is a means of stimulating him to make a more powerful resolution in favor of abstinence than any he has hitherto made; or it is just this once, and once doesn't count, etc., etc., *ad libitum*—it is, in fact, anything you like except *being a drunkard*. *That* is the conception that will not stay before the poor soul's attention. But if he once gets able to pick out that way of conceiving the various opportunities which occur, if through thick and thin he holds to it that this is being a drunkard and nothing else, he is not likely to remain one long. The effort by which he succeeds in keeping the right *name* unwaveringly present to his mind proves to be his saving moral act.[47]

This passage may be dismissed lightly as a piece of moralizing rhetoric, but we may also regard it as freighted with the ideas that made up the early cargo of James's nominalistic pragmatism. There is a footnote to this passage in which James refers us to Aristotle's discussion of incontinence and the doctrine of the practical syllogism in order to explain the perennial problem of why we so often see the better and do the worse. There are two kinds of premises, Aristotle's doctrine teaches, universal and particular. When we see the better we make use only of universals, for example, drinking too much is harmful; but in order to know what we must *do* in a concrete situation like that of the drunkard, it is necessary to have also particular premises, for example, I won't drink this because it will form a habit of drinking more than is good for me.

The "will to believe" doctrine of James is the doctrine of the practical syllogism generalized to cover metaphysical questions about God, freedom, and immortality, as well as moral situations. We know that Chauncey Wright, Peirce, and Holmes severely castigated James for confounding metaphysics with morals. James had first enunciated the doctrine as "the duty to believe" in conversation with these early intellectual companions of his in the Metaphysical Club in the 1870's, when agnosticism was riding high on the wave of evolutionism and seemed to leave morals without any firm foundation.

Chauncey Wright, about two months before his death, described his younger friend William James, in a letter to Grace Norton, as one "to whom the perfection of moral action and belief is in heroic conditions of life; and a creed adapted to these, however rare they may be in fact, is to him the true creed, covering the whole range of life, and prescribing a rule for the extremes of human action." [48] Then, after describing the debate between himself and the heroic young William, Wright reports that concessions were granted on both sides:

> I allowed that unproved beliefs, unfounded in evidence, were not only allowable, but were sometimes even *fit, becoming,* or appropriate to states of feeling or types of character which are deserving of approval, or even of honor. This fitness does not however amount to an obligation of duty. So far we are agreed, and he [James] retracts.[49]

James as a result changed the name of his doctrine to the "right to believe," but still fell afoul of the more implacable logical conscience of his friend Charles Peirce, who argued that the right *not* to believe in the absence of preponderant evidence was the *duty* of a man of science. Wright had been more understanding of James's moral point that a judgment *not* to hope and believe what accords with one's ideals signifies a betrayal of one's loyalty. On the question of the relation of metaphysics to morals, Wright took a more sociological and historical position than James: "The Nature still deserving our worship is the harmony of an elevated moral standard, pragmatically opposed to the claims of traditional institutions and sanctions . . . Natural rights were pragmatically real, so long as divine rights were.[50]

Returning to James's *Principles of Psychology*, we find the evolutionary approach of James continuing from his physiological analyses of instinct, emotions, and will to the intellectual operations of the mind. The laws of logic and logical order appear "in a mind which has come by some lucky variation to apprehend a series of more than two terms at once." [51] Experience itself is not a logical system, for it supplies the mind with the raw, unordered, primitive materials and problems which the intelligent animal has to put into order or suffer the consequences. Now, in scientific investigations systematic classification and the logical ordering of observed properties are the results of comparisons and functional correlations enabling us to predict and adjust ourselves to things to come. Why should the logical operations of the mind "form such a mighty engine for dealing with the facts of life?" James's answer to this epistemological puzzle of the ages begins with an instrumentalist view of concepts, and dates back to 1879, the year after the publication of Peirce's "How To Make Our Ideas Clear": "What is a *conception?* It is a *teleological instrument.* It is a partial aspect of a thing which *for our purpose* we regard as its essential aspect, as the representative of the entire thing." [52]

There are no privileged concepts or ways of conceiving things truer than any others; there are only more important ways, "more frequently serviceable ways." What James wished to repudiate as "utterly absurd" was the popular notion that "Science is forced on the mind *ab extra,* and that our interests have nothing to do with its constructions." [53] Helmholtz's account of the discovery of the law of the conservation of energy law is used by James as evidence of the scientist's sense of the plasticity of the conceptual materials in his hands, the *as if* character of the assumption of Nature's intelligibility in terms of physical constants. By translating the mixed-up order of sense-experience into ideal relations of kinds, number, form, equality, and so on, the scientific discoverer is able to verify "that the things of Nature turn out to act *as if* they *were* of the kind assumed." [54] James thought the scientist was thus obliged to declare that the experienced form of things is false and the ideal form true, "declarations which are justified by the appearance of new sensible experiences at just those times and places at which we logically infer that their ideal correlates ought to be." The test, then, of a scientific theory is the concrete sensible experiences predicted, for "thus the world grows more orderly and rational to the [scientific] mind." [55] If physics has been able to make progress by applying ideal systems of relations to the real world, so metaphysics ought to emulate physics, "simply confessing that hers is the longer task." [56] This conception of metaphysics as simply long-range science goes counter to the scholastic refutation of Darwinism.[57] The scholastic metaphysics arbitrarily legislates for all natural science, whereas philosophy, according to James in his *Principles of Psychology,* cannot gainsay scientific findings. Philosophy need not, however, stop short with the latter, James insists, for man has other interests beside that of science.

The upshot of James's Darwinism in psychology and his pragmatic philosophy was a profound conviction of the inconclusiveness of any and all scientific accounts of either the genesis or destiny of our mental lives.

The causes of our mental structure are doubtless natural, and connected, like all our other peculiarities, with those of our nervous structure. Our interests, our tendencies of attention, our motor impulses, the aesthetic, moral, and theoretic combinations we delight in, the extent of our power of apprehending schemes of relation, just like the elementary relations themselves, time, space, difference and similarity, and the elementary kinds of feeling, have all grown up in ways of which at present we can give no account. Even in the clearest parts of Psychology our insight is insignificant enough. And the more sincerely one seeks to trace the actual course of *psychogenesis,* the steps by which as a race we may have come by the peculiar mental attributes which we possess, the more clearly one perceives "the slowly gathering twilight close in utter night." [58]

This *Gotterdämmerung* of the natural science of psychology and of all that James had hoped to get from it for the alleviation of man's spiritual life must be ascribed in part to the growing *fin de siècle* sense among scientists that certainty was not attainable in natural sciences. The optimism of evolutionary determinists like Fiske had to go with the abandonment of certainty. There were three possibilities at the end of the last century for the philosophers who had pinned their hopes on the banner of evolution: one was to continue to hope that progress was inevitable but spiral, with occasional relapses; the second was to banish all ideals and values from science as beyond calculation or public confirmation; and the third was to admit the merely probable character of scientific laws, but to seek elsewhere in private experience the certainties and ideals no longer attainable by scientific method. William James turned to this third alternative when in his later metaphysical phase he stressed immediacy and the subjective method as more satisfying than the partial, piecemeal abstractions of science. It is no wonder that literary historians like Mr. Matthiessen have found a close bond between William and his brother Henry, the novelist, not only in their remarkable powers of expression but also in their extremely sensitive and broad vision of the minute and inexplicable complexity and subtle variety and range of feeling and thought whose elements required and yet eluded the most scrupulous and refined analyses. Henry, the literary psychologist, worshipped at the shrine of artistic perfection of form, while William turned from physiological psychology to metaphysical forms of fideism and intuition and even to the quest for a pragmatic proof that *inconclusiveness* and *indeterminateness* resided in the very bosom of thought and nature.

The heart of James's philosophy lies in his religious regard for the primacy of the rights of the individual. Since Darwin's idea of "spontaneous variations" seemed to James to lend scientific support to this central individualistic doctrine, it was eagerly embraced. And if any scientific theory failed to support the right and freedom of individuals to assert their moral natures, so much the worse for that scientific theory, for James sought the possibility of finding other grounds than that of neutral and hesitant scientific hypothesis on which to base our individual moral decisions and actions. In James's pragmatism, a scientific theory is limited in its meaning to the *individual* perceptions and experiences it leads to; and if in genuine, live, and momentous options logic does not lead to a clear indication of the preferable choice, because the evidence is divided evenly, why then should we not throw the whole weight of our passions and aspirations into the uncertain scales? We thus take the responsibility on ourselves for what is to happen to us, rather than let blind force or alien interests win the day. We lose by default if we stand paralyzed by hesitant, skeptical doubt and fear of uncertainty. The will to believe was James's answer to agnostic evolutionism and the dehumanized pseudoscientific quest for metaphysical certainty.

James did not extend his radical empiricism to social theory, not because he was a reactionary or rugged individualist, but because he looked to the individual rather than to economic, political, or legal institutions for social melioration. When he protested with indignation against our imperialism it was in the morally heroic spirit of a Whitman or Tolstoy, inspired by a profound love of freedom, but uninformed concerning the complex political and economic conditions and institutional changes required for its attainment.

James, Norton, and Abbot belonged to the New England Anti-Imperialist League formed in 1898 by a group of Boston liberals, who for five years expressed their indignation at our government's policies toward the Philippines, Haiti, and Venezuela. In his address to this league in 1903, James was extremely bitter and pessimistic about our government's national policy:

> The country has once and for all regurgitated the Declaration of Independence and the Farewell Address, and it won't swallow again immediately what it is so happy to have vomited up. It has come to a hiatus. It has deliberately pushed itself into the circle of international hatreds and joined the common pack of wolves. It relishes the attitude. We have thrown off our swaddling clothes, it thinks, and attained our majority. We are objects of fear to other lands. This makes of the old liberalism and the new liberalism of our country two discontinuous things. The older liberalism was in office, the new is in the opposition.[59]

The only hope for our future political evolution, added James, lies in abandoning the barefaced piracy of nationalistic imperialism and in making common moral cause with "the great international and cosmopolitan liberal party, the party of conscience and intelligence the world over . . . we are only its American section, carrying on the war against the powers of darkness here, and playing our part in the long, long campaign for truth and fair dealing which must go on in all the countries of the world until the end of time." [60]

In *Varieties of Religious Experience,* James compared the social ideas of the Utopian socialists with the Christian saints' belief in the kingdom of heaven. Four years before these Gifford Lectures, James had criticized Le Bon for condemning all forms of socialism as "crazy," and four years after the same lectures, he wrote to H. G. Wells with the warmest praise for his evolutionary socialism.[61] However, it was not the evolution of a scientific socialism that interested James (as it did Wells), but the possibility of emancipating the individuals whose spiritual lives were being thwarted by the growth of the greedy commercialism of the Gilded Age. Against it he vented his moral indignation in Tolstoyan fashion.

He exposed the cant and hypocrisy behind the pseudo-evolutionary arguments of social Darwinists who sought to justify imperialism and the exploitation of the weak. Though he was too ill-informed in political economy to criticize the economic theories of those who worshipped big business and the "bitch-goddess Success," he condemned them in no uncertain terms. His social and political liberalism belies the Marxist interpretation of pragmatism as the "reflection" and instrument of an exploiting class ideology. Though he was not the social engineer of American democracy, he will live as its moral prophet. His religious sense of the sacredness of the individual led him to regard most big institutions as corrupt: "We 'intellectuals' in America must all work to keep our precious birthright of individualism and freedom from these institutions. *Every* great institution is perforce a means of corruption—whatever good it may also do. Only in the free personal relation is its full ideality to be found." [62]

This was not a merely momentary outburst of passionate indignation. It was an expression of James's secular yet religiously democratic love of individual freedom which he had once called "empiricism's glory" in his famous address on "The Will To Believe." [63] In it James directed all his fire against agnostic evolutionism as "the queerest idol ever manufactured in the philosophic cave":

When I look at the religious question as it really puts itself to concrete men, and when I think of all the possibilities which both practically and theoretically it involves, then this command that we shall put a stopper on our heart, instincts and courage, and *wait*—acting of course meanwhile more or less as if religion were *not* true—till doomsday, or till such time as our intellect and senses working together may have raked in evidence enough—this command, I say, seems to me the queerest idol ever manufactured in the philosophic cave. Were we scholastic absolutists, there might be some excuse. But if we are empiricists, if we believe that no bell in us tolls to let us know for certain when truth is in our grasp, then it seems a piece of idle fantasticality to preach so solemnly our duty of waiting for the bell. Indeed we *may* wait if we will—I hope you do not think that I am denying that—but if we do so, we do so at our peril as much as if we believed. In either case we *act,* taking our life in our hands. No one of us ought to issue vetoes to the other, nor should we bandy words of abuse. We ought, on the contrary, delicately and profoundly, to respect one another's mental freedom: then only shall we have that spirit of inner tolerance without which all outer tolerance is soulless, and which is empiricism's glory; then only shall we live and let live, in speculative and practical things.[64]

This deep respect for one another's mental freedom—the glory of James's pragmatic empiricism—was appreciated by Peirce who had disagreed so radically with the will-to-believe doctrine of his lifelong friend:

His comprehension of men to the very core was most wonderful. Who, for example, could be of a nature so different from his as I? He so concrete, so living; I a mere table of contents, so abstract, a very snarl of twine. Yet in all my life I found scarce a soul that seemed to comprehend, naturally not my concepts, but the mainspring of my life better than he did. He was even greater in the practise than in the theory of psychology.[65]

CHAPTER SIX

Social Evolutionism: Fiske's Philosophy of History

1. JAMES VERSUS FISKE ON GREAT MEN IN HISTORY

The American representatives of Hegelian idealism—the St. Louis school of Louis Brokmeyer, Denton C. Snider, W. T. Harris, and the more formidable scholars Thomas Davidson and Josiah Royce—were more generous than James in appraising the evolutionary role of social institutions. Hegel had deified the state and endowed its laws with such omnipotent reason that in his political philosophy passive obedience became the primary virtue of the ordinary citizen. It is fair to infer that the American Hegelians (even Walt Whitman espoused the rhetoric of Hegel as a means of avoiding the impending clash between the North and the South), who had behind them more democratic traditions than the Prussian spokesman for the Almighty's cunning use of great men ("world-historical individuals" like Caesar, Napoleon, and Frederick the Great), ignored the Hegelian subordination of ordinary individuals to the inexorable dialectic of the unfolding of a Divine Plan. Early in his career at Harvard, James ran into one form of this idealism blended with Spencerian evolutionism in the philosophical historian John Fiske. He, like F. E. Abbot, we recall, "was sometimes present at meetings of the Metaphysical Club but held aloof his assent" to its pragmatic talk. Neither Fiske nor Abbot could look with favor at the skeptical turn the discussion of evolution frequently took. Fiske hoped that evolutionary science would become the handmaid of liberal Unitarianism rather than the instrument of enlightened skepticism. Abbot's more radical scientific theism was couched in Hegelian obscurities and never achieved the popular success of Fiske's eloquent and optimistic gospel of progress.

Fiske delivered a popular series of thirty-five lectures at

Harvard in 1872 on the Doctrine of Evolution, which were published two years later in a two-volume work, *Outlines of Cosmic
Philosophy.* We recall that James held a meeting in Cambridge in
1874—perhaps of the Metaphysical Club—at which he, Wright,
Peirce, Green, and Warner discussed the book with Fiske "who
fell asleep under our noses." Six years later, James and Fiske
were engaged in a controversy in the pages of the *Atlantic
Monthly* concerning the Spencerian philosophy of history and social evolution.

James's essay was entitled "Great Men, Great Thoughts, and
the Environment," and Fiske's rejoinder was "Sociology and Hero-
Worship, an Evolutionist's Reply to Dr. James." Their debate
sheds light on the impact of evolution on the growth of pragmatism in the social sciences in the United States after 1875.

Fiske was not one of the founders of pragmatism, though
there are elements in his thinking which he shared with his contemporary pragmatic friends at Cambridge: a liberal respect for
individual freedom and a faith that the growth of freedom could
be promoted in society by the progress and diffusion of science,
especially the theory of evolution. He differed from the early
pragmatists in clinging to Spencer's mechanical evolutionism,
though we shall see that he even gave that doctrine the richer
coloring of his own more liberal and more Christian idealism. He
is important to us for trying to do in his own way for American
history, anthropology, and political thought what Wright, Peirce,
and James were doing for physical sciences, logic, and psychology, namely, to emancipate them from traditional metaphysical dogmas. His early flirtation with the positivism of Buckle and
Comte gave way to an evolutionary philosophy of history and
theology. Though the founders of pragmatism did not accept his
Spencerian theology, they were concerned with the problems
he raised, and by disputing his evolutionary sociology and philosophy of history, they were stimulated to examine further the
social implications of evolution. The lawyer members of the Metaphysical Club were beginning to dig into the caked ground of
legal history, exposing the evolutionary fossilization of the common law, and preparing the soil for a revitalized, realistic, and
pragmatic theory of the law.

James opened his essay in the *Atlantic Monthly* by boldly

throwing a Darwinian bridge across the gap between the social and natural sciences: "A remarkable parallel, which to my knowledge has never been noticed, obtains between the facts of social evolution and the mental growth of the race, on the one hand, and of zoological evolution, as expounded by Darwin, on the other." [1] The parallel lies in the unpredictable occurrence of unusual persons and biological mutations, Darwin's "chance variations." There was no room or free play for such spontaneous individuals in Spencer's tidy, mechanical evolutionary scheme of nature or social history. The evolutionary philosophy of Spencer and of his disciples, John Fiske and Grant Allen, errs in explaining the complex processes of social evolution by means of an impersonal environmental determinism of physical, geographic, or ancestral conditions. You cannot explain, adds James, by way of illustration, what makes "the Harvard College of to-day so different from that of thirty years ago," [2] if you leave out the particular persons who make up the institution. The trouble with Spencerian evolutionism is that it tries to take the whole universe into its "impartial" account of human history. James objects, above all, to the Spencerian "adamantine fixity of a system of natural law":

> In the vagueness of this vast proposition we have lost all the concrete facts and links. And in all practical matters the concrete links are the only things of importance. The human mind is *essentially* partial. It can be efficient at all only by *picking out* what to attend to, and ignoring everything else,—by narrowing its point of view.[3]

The great merit of Darwin's theory is that he picked out two *relatively* independent phases or cycles of evolutionary change: (1) the production of new peculiarities by "the tendencies of spontaneous variation," and (2) the preservation or elimination of these variations by means of natural and sexual selection. Pre-Darwinian philosophers like Lamarck and Spencer had "all committed the blunder of clumping the two cycles of causation into one." [4] We know far too little about the invisible, molecular cycle of prenatal or congenital influences upon ova and embryos to be able to say how mutations are produced.[5] Once they are produced, the visible environment goes to work by natural selection to preserve or destroy. Now comes James's thesis:

The causes of production of great men lie in a sphere wholly inaccessible to the social philosopher. He must simply accept geniuses as data, just as Darwin accepts his spontaneous variations. For him, as for Darwin, the only problem is, these data being given, How does the environment affect them, and how do they affect the environment? Now, I affirm that the relation of the visible environment to the great man is in the main exactly what it is to the "variation" in the Darwinian philosophy.[6]

James admired Carlyle for his *Heroes and Hero-Worship,* and Gospel of Work.[7] The mutations of societies are due to the powerful initiative of leaders like Clive in India and Bismarck in Germany. Changes in intellectual history are also due to the bents and personal tones of exceptional minds. "The products of the mind with the determined aesthetic bent please or displease the community. We adopt Carlyle and grow unsentimental and serene. We are fascinated by Schopenhauer, and learn from him the true luxury of woe. The adopted bent becomes a ferment in the community, and alters its tone." [8]

Fiske in his rejoinder admits James is right in regarding "sociological variations" called geniuses as similar to Darwin's "spontaneous variations" in zoologic evolution, for both are unaccountable deviations from the average. But he saw no incompatible differences between Darwin's and Spencer's views of evolution, despite the fact that Wright, James, Peirce, and Darwin did. In order to bring geniuses under the Spencerian formula of evolution, Fiske defines a genius statistically: one who differs only in *degree* from most other individuals who all differ from one another in lesser degree. However, James would have been likely to reply, as Chauncey Wright once had indicated, that a large difference in degree was practically all one meant by a difference in kind. Fiske stuck to the point that all variations are subject to the selective action of the environment, as James admits; *ergo,* geniuses are also subject to the larger uniform laws of the environment. Furthermore, Fiske adds, Spencer does not deny the role of individuals provided that *due* reference is made to the surrounding social conditions. Fiske agrees with James that Grant Allen goes too far in ascribing all cultural changes to the geographical environment, and that Bagehot's psychological approach in *Physics and Politics* is much sounder. But the trouble with James's philosophy of history, Fiske properly insists, is that it does not do justice to the sociological role of institutions:

> The study of sociology, in short, is primarily concerned with *institutions* rather than with *individuals*. The sociologists need not undervalue in any way the efficiency of individual initiative in determining the course of history; but the kind of propositions he seeks to establish are general propositions, relating to the way in which masses of men act under given conditions.[9]

Fiske finally notes that Carlyle's and Froude's biographical method is inferior to Mommsen's comparative method and institutional approach to history; the great English historian Freeman, not less of a hero-worshipper than Carlyle, is also an institutional historian. He also notes that there has been an immense change in historiography between the time of Carlyle's and Macaulay's *biographical* histories (1840–50) and Freeman's *institutional* work (1880), owing to the evolutionary researches of Maine and Stubbs, Coulanges and Maurer. Fiske concluded his reply to James with the sharp rebuke that Freeman was "as ardent a hero-worshipper as Carlyle himself,—and vastly more intelligent." [10]

In the historiography of the period, Fiske stood opposed to those who like Goldwin Smith regarded free will as incompatible with a scientific approach to history. Fiske, in more modern fashion, cited Buckle's statistical method as leaving human volition and responsibility intact. Fiske, like Buckle, frequently fell back on the language of the "necessarians" to oppose uncaused, unconditioned free will (which he and Buckle found to be as meaningless as the fatalists' notion of unconditioned destiny); but it was clear to Fiske that the heart of the problem of free will is not the imputing of necessity or fortuitousness to events, but the pragmatic question of fixing responsibility in moral and legal situations relative to given psychological, social, and political conditions. History can scientifically ascertain what these conditions truly are, and thus serve as an indispensable auxiliary to the study of the problems of civilization. In any case, all the founders of pragmatism would agree with Fiske that unconditioned free will would render history a chaos: "The vast domain of History, numbering among its component divisions the phenomena of Language, Art, Religion, and Government, the products of social activity as well as the phases of social progress, becomes an unruly chaos, a Tohu-va-Bohn, where event stumbles after event, and change jostles change, without sequence and without law." [11]

Fiske agreed with Buckle, Mill, and James that free will was subject to empirical and not metaphysical conditions, and that the metaphysicians' concept of free will ought not "deter us from applying scientific methods of interpretation to the phenomena of human history." [12] However, like Buckle and Mill, Fiske regarded all empirical phenomena, including those of social history, as conforming to fixed laws, so that all societies necessarily pass through the same stages of development. It is only the limitation of our intellectual vision that prevents us from ascertaining with certainty the immutable laws governing the complex and diverse phenomena of society and history.[13] Froude's arguments against a science of history are simply an assertion of the complexity of historical fact and not a proof of the impossibility of scientific investigation of history. Against Froude's argument that historical phenomena like the founding of new religions are unpredictable, Fiske maintained, "What could not have been predicted was the peculiar character impressed upon these [religious] movements by the gigantic personalities of such men as Mohammed and Omar, Sakyamuni, Jesus and Paul. What could have been predicted was the general character and direction of movements." [14]

This is an important defensible view of the possibility of scientific historiography, for the methodological principle used by Fiske may be said to conform to the presuppositions of the method of science in all fields. That is, we do not in cognition *exhaust* the *individual* character of events, but only apprehend their communicable general features. Thus, individuality and law are historically compatible. An act of free will is, for Fiske, an individuation of equally determined possibilities. There is no transcendental "Will" which supersedes natural law. In this connection Fiske adheres to his youthful positivism; he regards the metaphysical problem of free will as a pseudo-problem generated by "confused and inaccurate verbiage."

Strip the question of the peculiar metaphysical jargon in which it is usually propounded, restate it in very precise scientific language, and it becomes a very easy question to answer. Would that science presented none more difficult! Confused and inaccurate verbiage is responsible for the chronic disputation upon this subject. Nowhere else is Berkeley's complaint so thoroughly applicable, that in dealing with metaphysics men first kick up a dust and then wonder why they cannot see through it.[15]

But neither Berkeley nor Fiske actually succeeded in ridding themselves of metaphysics; the good bishop's philosophy rests on a theory that knowledge and reality are akin in spiritual substance, and Fiske insisted that evolution supported a cosmic and Christian theism. Yet the illusory belief that the onward and upward process of evolution was a purely scientific and inspiring conclusion led Fiske into a quasi-positivistic and quasi-pragmatic position with respect to traditional metaphysics. He espoused Bain's doctrine that the will is simply the most powerful among a conflicting set of emotional tendencies to act one way rather than another.[16]

American historiography owes something to Fiske for putting it on a more "scientific" basis than it had enjoyed previously. His works in American history have been superseded, and scientific method in historical research is nowadays taken for granted without making a fetish of it.[17] But he cannot be neglected by the historian who wishes to understand the emergence of pragmatic liberalism in social sciences out of theological and metaphysical attempts to embalm them for eternity. Fiske believed that he had effected the reconciliation of science and religion in his *Outlines of Cosmic Philosophy, Based on the Doctrine of Evolution, with Criticisms on the Positive Philosophy* (1874), *The Destiny of Man, viewed in the Light of His Origin* (1884), and *The Idea of God as affected by Modern Knowledge* (1885). After 1885 he turned from cosmic theism to American politics and history, subjects that were growing in popular interest with America's rapid industrial, westward and colonial expansion. He made good use of his linguistic precocity to delve into the Spanish, Italian, and French sources of the discovery and colonization of America. We shall not go into the dozen volumes he wrote on American history, but confine ourselves to the earlier cosmological and anthropological evolutionism of Fiske. His views were well known to his pragmatic friends at Cambridge who shared with him a philosophical and humanistic interest in the scientific progress of the times, and were all impressed with the scope of Comte's, Buckle's, Maine's, and Spencer's sociological syntheses. Whoever wishes to learn something of the ways these European positivistic philosophies affected American thought during the formative years of pragmatism will do well to look into the mirror of Fiske's encyclopedic mind.

2. THE POSITIVISTIC SOURCES AND IDEALISTIC OUTCOME OF FISKE'S EVOLUTIONISM

When he was only eighteen years old, Fiske wrote his mother that he considered it his "duty to mankind as a Positivist to subscribe [to Herbert Spencer's work]; and if I had $2,000,000. I would lay $1,000,000. at Mr. Spencer's feet to help him execute this great work." [18] In these youthful letters home, Fiske exuberantly longs for the works of all the English Positivists, J. S. Mill, George H. Lewes, John Grote, Thomas Buckle, Sir John Herschel, Spencer, Alexander Bain, Robert Mackay, Charles Darwin, and Sir Charles Lyell. In his senior year at Harvard he moved next door to Professor Gurney's with the positivistic Chauncey Wright and Frothingham from Brooklyn living on the floor above. [19]

That Fiske had a second-hand knowledge of physical sciences as well as an unduly high estimate of Spencer's attainments as a scientist is revealed by his naïve statement that Spencer's essay on the nebular hypothesis (in the *Westminster Review* of July 1858) "is the greatest production of the human intellect since the "Principia" of Newton. With Laplace's own data he proves what Laplace couldn't." [20]

In another early article in the *North American Review,* Fiske attempted again to apply the principles of evolution to some of the problems of explaining the growth of modern languages from the variations of local dialects. [21] In this field Fiske was much more at home; he had early in his high school days distinguished himself as a gifted linguistic scholar. Fiske's humanistic positivism emerged from his scholarly precocity in language and history *before* he came under the spell of Comte's and Spencer's syntheses of natural and social sciences. "If I may cite my own experience, it was largely the absorbing and over-mastering passion for the study of history that first led me to study evolution in order to obtain a correct method." [22]

His first youthful essay on "Buckle's Theory of Civilization" (1861) and subsequent criticisms of Comte's positivism in favor of Spencer's evolutionism led to his appointment in 1869 by President Eliot of Harvard [23] to a university lectureship on positivism. He became instructor in History the next year, and served for seven years (1872–1879) as assistant librarian of the university.

Buckle's positivism seemed to Fiske too physicalistic, and Comte's too rigid, especially in its last theocratic phase, until he thought he found in Spencer's evolutionism the ideal system for coördinating man's material and spiritual interests. From our present-day perspective, the history of positivism shows two divergent developments, one stemming from the physical sciences (Mach, Helmholtz, Kirchhoff, Wright, Stallo), and the other from the less exact biological and social sciences (Buckle, Taine, Comte, Huxley, Mill). Evolutionary positivists like Spencer and Fiske assumed that physics gave us final certainty concerning the laws of nature; they sought to establish the same sort of finality in the organic and social ("super-organic") worlds. Evolution seemed to them to participate in the mechanical determinism which the physical sciences had established in the laws of gravitation and conservation of energy. The grand aim of Fiske was to find a cosmology which would subsume physical, biological, psychological, and social phenomena under one system of laws, and at the same time guarantee human progress.

While waiting for clients that did not come to his Boston law office, Fiske went into rapture reading Sir Henry Sumner Maine's *Ancient Law,* and called it "a new Epoch of my Life." [24] Here was a work that superseded his earlier positivistic admiration of Comte and Buckle, for it gave evidence of man's moral as well as intellectual progress (Buckle having denied moral progress) in evolutionary terms. There is no doubt that Fiske substituted the immutable laws of astronomical and biological evolution for the eternal verities of revealed religion, a transfer of faith which Comte's and Buckle's positivism failed to accomplish, to Fiske's optimistic mind, as effectively as Spencer's metaphysical system.

When Fiske sent Darwin a copy of his *Outlines of Cosmic Philosophy* he did not seem to realize the great difference in *method* that separated Darwin's from Spencer's study of evolution. Darwin, who generally disliked personal controversy, was tactful in impressing this difference on Fiske:

I wish some chemist would attempt to ascertain the result of the cooling of heated gases of the proper kinds in relation to your hypothesis of the origin of living matter. It pleased me to find that here and there I had arrived from my own crude thoughts at some of the same conclusions with you; though I could seldom or never have given my reasons for such conclusions. I find that my mind is so fixed by the inductive method that I cannot appreciate deductive reasoning: I must begin with a good body of facts and not from a principle (in which I always suspect some fallacy) and then as much deduction as you please.

This may be very narrow-minded; but the result is that such parts of H. Spencer as I have read with care impress my mind with the idea of his inexhaustible wealth of suggestion, but never convince me; and so I find it with some others. I believe the cause to lie in the frequency with which I have found first-formed theories to be erroneous.[25]

Fiske's friend and biographer, John Spencer Clark, insists (too much, I think) that the main object of Fiske's *Outlines of Cosmic Philosophy* was antipositivistic:

nothing less than the freeing of the doctrine of Evolution from all kinship with the Positive Philosophy of Auguste Comte, from all identification with atheism or materialism, while at the same time rounding it out into a philosophic system based on science; a system consisting of affirmations as to the existence of Deity, accompanied by verifiable data regarding the cosmic universe, with man's place in it with his rational mind, as a unified, ever-developing manifestation of Deity. And it was for the completion of this important task that he desired converse with Spencer, Darwin, Huxley, Lewes, Tyndall, Hooker, Clifford, Lockyer, and a few others of the new school of scientific thought in England.[26]

However, the main work of Fiske was at least consciously forged with the aid of positivistic criticism, even though the outcome was highly spiritualistic. For example, Fiske reported that W. K. Clifford went to work "by punching through about six pages of my Nebular Hypothesis at once, and so saved me from getting into trouble hereafter." [27]

Fiske was unable to go along with Clifford, who doubted

the classical view, that persisted in Maxwell, that "the laws of Mechanics are absolutely the same throughout eternity." [28] The reason for Fiske's disagreement with Clifford lies in the fact that the harmony of science and religion in Fiske's mind depended on the recognition of eternal principles at work in *both* the physical and spiritual worlds. The distinction between matter and thought is, however, as absolute as the laws governing both. Yet Fiske regarded both psychological and physical laws as phenomenal: "In science we make no use of the conception of a spiritual substance (or of a material substance, either) because we can get along sufficiently well by dealing solely with qualities." [29]

There is the same phenomenal theory of qualities in Chauncey Wright's doctrine of the neutrality of scientific method and in many of William James's statements in the *Principles of Psychology*. However, Fiske was less consistent than either of his two Cambridge friends on this point; for he declared himself in agreement with Spinoza on the infinite power of thought,[30] and in defending Spencer against the charge of materialism, he declared that Spencer's evolutionary *Principles of Psychology* did not dispense with God and immortality, as Haeckel's and Büchner's materialistic systems did. He had even obtained Spencer's approval to emend a passage in quoting from Spencer's work (I, 158) by changing the word "nervous" to "psychical." In line with his own Christian idealism, Fiske was quite sure that the progress of science was not going to eliminate the religious sentiment but would guide it as "emotional prompting toward completeness of life." [31]

> Religious feeling has survived the heliocentric theory and the discoveries of geologists; and it will be none the worse for the establishment of Darwinism . . . One good result of the doctrine of evolution, which is now [1875] gaining sway in all departments of thought, is the lesson that all our opinions must be held subject to continual revision, and that with none of them can our religious interests be regarded as irretrievably implicated.[32]

Now a phenomenalistic view of science is not incompatible with a metaphysical theory of allegedly higher realities of a spiritual sort transcending scientific knowledge. Fiske, however, oscillates between two incompatible theories of science: one, along

Hume's lines, radically empiricistic and skeptical of all claims to
knowledge beyond what is sensibly verifiable; the other, along
Kantian lines, which restricted science to "phenomena" but which
permitted reality to extend beyond objects of possible sensible
experience to a supersensuous world of "unknowable" noumena.
The latter world leaves room for religious faith, and, in the end,
that is what Fiske wished to safeguard and to substantiate by interpreting evolution as aiming at a religious goal. What is left
unclear is whether scientific method can verify the existence of
such an evolutionary goal or simply say nothing to contravene
one's faith in it as an unknowable noumenon. The logical difficulty of knowing the unknowable destiny of man is uneasily
skirted by Fiske's theistic version of Spencer's Unknowable. What
Fiske did, in order to save religious faith, was to transfer metaphysical *certainty* to the scientific law of evolution. This procedure, however, implies a tacit admission that *certainty* is not attainable by metaphysical methods. Thus he never abandoned
his early positivistic doubts about metaphysics.

These earlier radical positivistic doubts are found throughout the *Outlines of Cosmic Philosophy* in such passages as the
following: "The truth of any proposition, for scientific purposes,
is determined by its agreement with observed phenomena, *and
not by its congruity with some assumed metaphysical basis.*" [33]
Copernicus and Newton are cited as in agreement with this view.
Copernicus is quoted: "Neque enim necesse est eas hypotheses
esse veras, imo ne verisimile quidem, sed sufficit hoc unum, si
calculum observationibus congruentem exhibeant." [34] And Newton is credited with having "accurately discriminated between the
requirements of science and the requirements of metaphysics"
and for having clearly seen in anticipation of Kant that "while
metaphysics is satisfied with nothing short of absolute subjective
congruity, it is quite enough for a scientific hypothesis that it
gives a correct description of the observed coexistences and sequences among phenomena." [35] Newton's "hypothesis" of universal gravitation, Fiske tells us, did not assert a metaphysical
cause for the action of matter on matter at a distance. "All that
the hypothesis really asserts is that matter, in the presence of
other matter, will alter its space-relations in a specified way;
and there is no reference whatever to any metaphysical *occulta*

vis which passes from matter in one place to matter in another place." [36]

Scientists do employ "fictions" like "attraction" and "infinitesimals" and "atoms." Lewes had shown the legitimacy of these "scientific fictions" in the domain of verifiable coexistences and sequences of phenomena, and Fiske accepted his theory. Here is a significant link in the history of the evolution of pragmatism, which makes much of the theory of fictions as working hypotheses.

The law of evolution, claims Fiske, has the same universality as the law of gravitation in relation to our experience.[37] Thanks to Spencer, the prophetic dream of Bacon of philosophy as an organism of which the various sciences are members has been realized, and the universe being thus shown inductively to be a cosmos rather than a chaos, the true philosophy is properly named the cosmic philosophy.

3. SOCIAL EVOLUTION AND THE DESTINY OF MAN

Comte, Lewes, and Mill by stressing the *intellectual* stages of social development ignored the nonrational or emotional factors in the evolution of man. Spencer came closer to the truth when he said that "Ideas do not govern the world; the world is governed by feelings, to which ideas serve only as guides." For Fiske the law of social evolution involves progress from anthropomorphism (covered by Comte's first two stages of theology and metaphysics) to "Cosmism" rather than to Comte's third stage, "Positivism." "Cosmism" was intended by Fiske to include both a moral and theological passage from egoism to altruism, from self-protection under a tribal military leadership thriving on war and conquest to love and protection of all mankind under an international order of coöperation. This millennium of world peace would come "probably at no very distant period"—so great is Fiske's evolutionary optimism in the early 1870's—"when warfare shall have become extinct in all the civilized portions of the globe." [38] However, Bagehot's *Physics and Politics* (1872) is praised by Fiske for showing that prior to any millennium of world peace, evolution requires the increased military power of the most civilized nations in order to develop in their people the highest character for cultural world leadership. In this theory of

civilization, Bagehot and Fiske assume the now dubious Lamarckian theory of the inheritance or slow transmission of socially acquired traits. "Men's natures have, through long ages of social discipline, become in some degree adapted to the social state . . .
Modern society can count upon an organic or instinctive conform-
ity to law on the part of individuals, upon which ancient society could not count." [39]

Social evolution aims primarily at a *moral* goal in Fiske's
philosophy of history. Comte lacked the *general* idea of a *science
of genesis* and could not have established the evolutionary law
of social progress before the 1870's,[40] though Comte's law of the
three stages is a first approximation. Von Baer's great treatise on
embryology in 1829 applied the idea of the science of genesis to
organic phenomena, but the idea was too novel for Comte, especially in the latter part of his life when he ignored the then growing scientific literature. On Comte's death in 1857, Spencer's essay
"Progress: Its Law and Cause"

first definitely extended the law of organic development to historic phenomena; although he had ever since 1851 been visibly working toward that
result, and had in 1855 reached that grand generalization of the development of both life and intelligence, regarded as processes of adjustment,
which underlies the law of social progress here expounded. It was this
splended series of researches, culminating in the announcement of the universal law of evolution, in 1861, which supplied a new basis for all the
sciences which treat of genesis. And finally, in 1861, the further clue to
these special laws was given by Sir Henry Sumner Maine, whose immortal
treatise on *Ancient Law* threw an entirely new light upon the primitive
structure of society.[41]

Fiske is well aware that the notion that society is an organism, as Plato imagined in his *Republic* and Spencer in his
sociology, is merely an analogy.[42] Even Spencer realized the
weakness of the analogy when he, as a rugged Anglo-Saxon individualist, maintained that the living parts of a society "do not
and cannot lose individual consciousness," and that "the community as a whole has no corporate consciousness. The corporate
life must here be subservient to the lives of the parts; instead of
the lives of the parts being subservient to the corporate life." [43]

This anti-Hegelian view of the individual's relation to society is, however, marred by a logical failing shared by the a

priori thinking of "German Darwinism," namely, the assumption
that the individual is enmeshed in an inexorable law of social development that progresses by *necessary* stages. Instead of drawing these necessary laws of social history from the unfolding inner nature of an unverifiable absolute mind, Fiske looks to the
more exact physical sciences for his model of evolutionary necessity: "As surely as the astronomer can predict the future state of
the heavens, the sociologist can foresee that the process of adaptation must go on until in a remote future it comes to an end in
proximate equilibrium." [44]

What is this social goal of "proximate equilibrium," a phrase
which is borrowed from physical science but is intended to con-
vey a moral ideal? Fiske's answer is admirable in its ethical idealism and world-embracing vision; but is far from being logically
derived from physical considerations, for it is inspired by a humanitarian hope of world federation and peace.

> The increasing independence of human interests must eventually go
> far to realize the dream of the philosophic poet, of a Parliament of Man,
> a Federation of the World.
> When the kindly earth shall slumber, lapt in universal law, and when
> the desires of each individual shall be in proximate equilibrium with the
> simultaneous desires of all surrounding individuals. Such a state implies at
> once the highest possible individuation and the highest possible integration
> among the units of the community; and it is the ideal goal of intellectual
> and moral progress. [45]

Richard Hofstadter, in a comment on Fiske as a synthe-
sizer of evolutionism and Anglo-Saxon expansionism, attributes
to him "a bumptious doctrine of racial destiny" based on his
long-held belief in "Aryan race superiority." [46] This is not quite
fair to Fiske's evolutionism, for Hofstadter does not indicate that
Fiske's racialism is more cultural than biological (except, perhaps, through Lamarckian inheritance); and secondly, that Fiske
departs widely from Spencer's laissez-faire individualism and imperialism in their respective theories of the role of economic competition and Anglo-Saxon expansionism. Finally, never did Spencer praise local self-government and federal democracy as Fiske
did in his lectures on American political ideas (1880). The advantages of world peace and federation for civilization are much
more strongly pointed up by Fiske than by Spencer, though it is

likely that Fiske's lecture on manifest destiny (1885), empha-
sizing the cultural hegemony of the English-speaking nations
toward the end of the century, played into the hands of more
militant British and American imperialists. In any case, there is
no question but that Fiske thought that political evolution provided a scientific basis for historical and social progress approaching ultimately the Christian pacific ideal of brotherly love and
coöperation. In his *Destiny of Man* Fiske traced three methods of
political development: (1) conquest without incorporation, yielding the despotic enslavement of conquered people, as in the case
of the ancient Persians, (2) conquest with incorporation, as in
the case of the Roman empire whose main weakness was the failure to devise an adequate democratic method of representation,
and (3) the "highest method of forming great political bodies,"
namely, that of *federation*. "The element of fighting *was* essential
in the two lower methods, but in this [third method] it *is* not
essential." [47] The shift in tense indicates precisely wherein Fiske's
evolutionism differs from Spencer's more static and rugged individualism. In these pages Fiske anticipated his British friend and
the greatest defender of Darwin, Thomas Huxley, who was later
to express the same cosmic evolutionary struggle between the
predatory and ethical nature of man.

The evolution of nationalities has progressed from the physical community of tribal or clannish descent to the psychical community of interests. These interests must, if there is to be social
progress rather than stagnation, not form into what Bagehot
called the "cake of custom" in which individual variations are
immobilized. The cementing and breaking of the cake of custom
must go on simultaneously rather than successively, as Bagehot
supposed when he set "the age of discussion" in the most recent
period of civilization. The English have made such great progress
because they were made up of Angles, Saxons, Danes, and Normans who, like their later pioneer Yankee descendants, broke the
cake of custom of their homeland and emigrated to new territory,
while the Orientals have accepted submissively the tyranny of
their customs and immobility.

Fiske's Anglo-Saxon cultural predilections are apparent in
his statement that "no previous century ever saw anything approaching to the increase in social complexity which has been

wrought in America and England since 1789." [48] Since increase in complexity automatically entailed progress, it is no wonder then that Fiske was so popular a lecturer in England as well as in the United States.

The evolutionary process of history was turned on itself when Fiske attempted to interpret all history in evolutionary terms. Royce thought—though Fiske disagreed—that Spencer was ignorant of the historical aspect of his evolutionary ideas. In any case, there is no doubt that Spencer welcomed the support of his American disciple, Fiske the historian. Spencer adhered closely to the deductive method of rational mechanics on which he was nurtured as an engineering student. But the rugged laissez-faire individualism that Spencer's evolutionism promoted in the ethical and economic world was far from the humanitarian or refined individualism that James, Wright, and Fiske himself envisaged as the worthy outcome of social evolution. Furthermore, Fiske was much more democratic in his reasons for favoring Spencer's disapproval of centralized government, for Spencer feared governmental control of private enterprise in the form of pro-labor reforms or public relief, as contrary to the law of the survival of the fittest, whereas Fiske pointed to the American tradition of town-hall democracy and to the growth of philanthropy as a sign of an evolving humaneness in political history. Finally, Fiske had a Unitarian's faith in the eventual brotherhood of man, good will and peace on earth, and transfigured Spencer's "Unknowable" by endowing it with Christian significance far from discernible in Spencer's dry agnostic formulas. What made Fiske's version of Spencer so popular in the United States was due more to Fiske's New England transcendentalist vision of the coming millennium than to his allegedly scientific defense of evolution ("the cosmic vehicle for the second coming of Christ"), though there is no doubt that the increasing prestige of Darwinism aided considerably in giving Fiske a large sympathetic audience. He was even accorded recognition as having made an "original" scientific contribution to the theory of evolution by virtue of his theory of the importance of the prolonged infancy of man. That theory, which was not so original as Fiske thought, had a special appeal to Americans conscious of the infant culture of their country as promising a greater future than the older cultures of Europe encased in Bagehot's "cake of custom" and impervious to new ideas. The American scholar had not yet fulfilled Emerson's declaration of self-reliance and independence of European models.

When he sent an outline of his views on prolonged infancy to Darwin, he met with some encouragement in Darwin's reply. The idea of prolonged infancy, and the lengthening care and education of children implied by it, was used by Fiske to explain: first, the differentiation of primitive herd-life into family groups; secondly, the possibility for each new generation to improve cumulatively upon the ideas and customs of its predecessors; finally, prolonged infancy made possible the growth of altruism.[49]

Fiske attached such great importance to his evolutionary theory of the role of prolonged infancy that he went so far as to regard it as "the whole explanation of the moral and intellectual superiority of men over dumb animals." [50] The moral development of man he associated with the development of the family, and consequently, with the growth of altruism. Fiske associated man's intellectual powers with "mental plasticity, or ability to adopt new methods and strike out into new paths of thought." These characters of plasticity and adaptability and innovating creativity are not entirely due to evolutionary theory, for we find them as features of the mind in Locke's *tabula rasa*, in epistemological theories of the practical utility of perception, and in Kant's and the German Romantic philosophers' theory of "productive imagination." But Fiske used these ideas to support his evolutionism which included the idea of social progress, notwithstanding Wright's, Peirce's, and James's criticisms of the extension of mechanical science to the analysis of mind. Fiske was even willing to speculate with the idea of a gradual acquisition of immortality as a "transcendent" consequence of his theory of prolonged infancy: "Though the question is confessedly beyond the reach of science, may we not hold that civilized man, the creature of an infinite past, is the child of eternity, maturing for an inheritance of immortal life?" [51]

Fiske adopted an evolutionary approach to ethnological questions at the very outset of his discussion of the aboriginal peoples of ancient America, not hesitating to draw upon biological evolution to explain ethnological distribution. The great differences between the cultures of diverse Indian tribes in both Americas do not, he insists, preclude a common racial stock. He attacks "the vague notion that grades of culture have some necessary connection with likenesses and differences of race. There is no such necessary connection." [52] Sir John Lubbock had in his evolutionary *Origins of Civilization* (p. 11) pointed out that "different races in similar stages of development often present more features of resemblance to one another than the same race does to itself in different stages of its history," and Fiske concurs completely in this important principle of cultural history and ethnology.

The influence on Fiske of Lewis Morgan, the anthropologist from Rochester, who, in his work on *Ancient Society* (1877), distinguished various stages of savagery, barbarism, and civilization, is very marked. Morgan regarded the making of pottery as the criterion for distinguishing the barbarian from the savage, for pottery along with the domestication of animals presupposes village life, which in its simple agricultural type of living is the mark of "barbarism." There are, in Morgan's analysis of cultures, three subdivisions, "low, middle, and upper statuses" of each of the three stages of savagery, barbarism, and civilization; but Fiske warns us against accepting too simple a classification. Morgan's "brilliant classification of the stages of early culture will be found very helpful if we only keep in mind the fact that in all wide generalizations of this sort the case is liable to be somewhat unduly simplified." [53] This caution regarding Morgan's analysis shows Fiske to be a more discerning and critical mind than Engels, who eagerly and uncritically adopted Morgan's stages because they fitted into his dialectical materialism.

That Morgan's useful work was based on his study of American Indian anthropology was regarded by Fiske as important for showing how an evolutionary study of aboriginal cultures would lead to a reëvaluation of early European civilization.

It was the study of prehistoric Europe and of early Aryan institutions that led me by a natural sequence to the study of aboriginal America. In 1869 after sketching a plan of a book on our Aryan forefathers, I was turned aside for five years by writing "Cosmic Philosophy." During that interval I also wrote "Myths and Myth-Makers" as a side-work to the projected book on the Aryans, and as soon as the excursion into the field of general philosophy was ended, in 1874, the work on that book was resumed.[54]

4. EVOLUTION VERSUS REVOLUTION

Fiske's contemporary biographer, T. Sargent Perry writes:

> The evolutionary theory, after making its way into physical science, had taken possession of the history of politics, of literature, of art, and finally was uniting even with its old foe, theology, in the study of early religions and the investigation of the Biblical books. While it had thus taken hold of advanced thought, it had also become known to the general public, who no longer feared its methods. To this agreeable harmony Fiske greatly contributed.[55]

Fiske aimed to extend the harmony portended by his cosmic doctrine of evolution into religion and politics at a time when both were undergoing the revolutionary travails of "the higher criticism" of Christianity and the social upheaval signalized by the Paris Commune of 1871. The last chapter of his chief philosophic work, *Outlines of Cosmic Philosophy*,[56] reveals an evangelical faith in an evolutionism that was to fuse science and the Protestant tradition of individualism against the "static" claims of the revival of Catholicism, of the Comtean priesthood of positivistic sociologists, and of the more radical Jacobin atheism of the French communists.[57] By making clever use of historical relativism, Fiske is able to play his evolutionary doctrine against both reactionary and radical extremes of religion and politics which ignore the *irreversible, cumulative,* and *gradual* processes of evolution. Catholics like De Maistre in France[58] and Mivart in England fail to realize that civilization cannot go back to a medieval, feudal, static society, though it is understandable why there was a Catholic reaction to the excesses of the French Revolution and to the attempts of Voltaire and Rousseau ("this greatest of Sophists")[59] to scrap orthodoxy overnight. Mivart, whom Wright had criticized for his anti-Darwinism in biology, is taken to task by Fiske for having written that the Darwinian doctrine of evolution tended to degrade man intellectually and morally and lead to "horrors worse than those of the Parisian Commune."[60] Not that Fiske did not share Mivart's sense of the "horribleness" of the revolutionary outbreak of the Parisian communards; but it was blaming evolutionary science for it that irritated Fiske. Instead of taking Wright's position of agnostic neutrality against Mivart, Fiske argued that *both* the Catholics and the communists were *unhistorical* in their political philosophies. Men's actions are the outcome of their individual habits and characters, and these are formed by slow, evolutionary processes that are irreversible and cumulative, so that we cannot either go back to the Middle Ages or change society by violent revolutionary methods. Social reform is needed but can come about only through gradual scientific enlightenment and not by recourse to medieval authority or violence. Scientific progress is itself the outcome of a cumulative evolution, and the continuity of evolution cannot brook either miracles (creation *ex nihil*) or revolutionary violence.

There is a modern touch to some of Fiske's impassioned criticisms of the communistic theories which would suppress individual freedom in science and society; the communists would "regulate human concerns by *status* and not by contract," which would completely invert what Sir Henry Maine had shown to be the evolution of society from the primitive communism of property and wives.

> To crush out capital and with it the possibility of any industrial integration, to abolish the incentives which make man sow today that he may reap in the future, to destroy social differentiation by constraining all persons alike to manual labor, to strangle intellectual progress by permitting scientific inquiry only to such as might succeed in convincing a committee of ignorant workmen that their discoveries were likely to be practically useful, to smother all individualism under a social tyranny more absolute than the Hindu despotism of caste; this desire, it is obvious, is simply the abnormal desire to undo every one of the things in the doing of which we have seen that social evolution consists.[61]

What Fiske did not realize was that the Marxists themselves appealed to an evolutionary philosophy of history, and even to Darwin, to support their revolutionary political program. Though they called themselves scientific historians and materialists, they shared with Fiske the Hegelian belief in a single inexorable logic of necessary stages of historical development. However, Fiske pinned the goal of evolution to individual moral and religious realization of human aspirations, whereas the Marxists looked to the transformation of social institutions by the material forces of production. But Fiske's fears of the threatened loss of individual rights of freedom, of scientific inquiry, and of civil liberties remain well grounded, at least, as a major problem of our own age of social planning and of liberalism struggling against totalitarianism.

In religious matters, the liberal Fiske was willing even to leave orthodoxy alone rather than supplant it suddenly with no "new system of conceptions equally adapted to furnish general principles of action . . . The evolutionist, therefore, believing that faith in some controlling ideal is essential to right living, and that even an unscientific faith is infinitely better than aimless scepticism, does not go about pointing out to the orthodox the inconsistencies which he discerns in their systems of beliefs." [62]

The same evolutionary gradualness of change of men's opinions and feelings applies to politics, which ultimately depends for reforms on the integrity of the individual's character. Fiske was the only member of our group to deal at length with political problems, but they all shared with him a profound historical sense of the impracticability of sudden reforms, and a high ethical sense of the *individual's* responsibilities in a progressive democracy.

> Since it is the plain teaching of history that the group of institutions making up the framework of society at any given period cannot be violently altered without entailing a partial disintegration of society; since any custom or observance can be safely discontinued only when the community has grown to the perception of its uselessness or absurdity; and, above all, since the integrity of society depends in an ultimate analysis, not upon its institutions . . . but upon the integrity of its individual members; it follows that the evolutionist will look askance at the panaceas of radical world-menders, refusing to believe that the millennium can be coaxed or cheated into existence until men have learned, one and all, each for himself, to live rightly. The only utopian ideal which he can consistently cherish is that of contributing his individual share of effort to the improvement of mankind by leading an upright life, and applying the principles of common-sense and of the highest ethics within his ken to whatever political and social questions may directly concern him as a member of a progressive community.[63]

Fiske seems to have become more idealistic and conservative as he grew older. His crusading synthesis of evolutionary science and Christian idealism never brought him to anything as radical as Francis Ellingwood Abbot's scientific theism. Nor did his theory of the inevitability of gradual progress through long spans of historical evolution lead him to be as critical of existing laws and institutions as his Cambridge friends. A more realistic theory of the evolution of the particular social institution of the law appears in the legal philosophy of his neighbor, Nicholas St. John Green.

CHAPTER SEVEN

The Pragmatic Legal Philosophy
of Nicholas St. John Green

Peirce and James, under the influence of Chauncey Wright, insisted that pragmatism would clarify metaphysical disputes about the meanings of ideas. Peirce said of pragmatism: "It is not to philosophy only that it is applicable." Since metaphysical ideas about causation, free will, mind, and moral responsibility still permeated the social sciences, it is significant historically to note that there were more lawyers than natural scientists in the Metaphysical Club where, Peirce alleged, "pragmatism saw the light of day." Wright, Peirce, and James were trained in the natural sciences. Green, Fiske, Holmes, and Warner were graduates of the Harvard Law School, practiced law, and, excepting Fiske, with brilliant success. One of them, Holmes, was to attain the highest office and distinction of the American bench. We owe to these lawyers—so far as the history of American social thought goes—the first major steps in the extension of scientific thinking to one of the most difficult of social disciplines, the clarification of the nature and sources of positive law. While only those trained in the law can appreciate and appraise their strictly legal work, there is much of larger historical and philosophical interest for us in their lives and in the spirit of their writings to justify Holmes's inspired dictum that "a man may live greatly in the law as well as elsewhere; that there as well as elsewhere his thought may find its unity in an infinite perspective; that there as well as elsewhere he may wreak himself upon life, may drink the bitter cup of heroism, may wear his heart out after the unattainable." [1]

I am not concerned (nor am I competent) to discuss the legal details of the writings of the philosophical lawyers connected with the rise of pragmatism. What I wish to exhume is the pragmatic social philosophy underlying their legal labors, and to show how it is related to their interpretations of evolution in human history. Despite the diverse versions of evolution held by the members of the Metaphysical Club, they shared a functional conception of institutions like the law, viewing them as the cumulative social product of practical decisions. They aimed to make our social thinking flexible, humane, and scientific, to cut into the cake of custom, though they were well aware of the crusty hardness of tradition and authority resistant to change. As the natural scientists were revising their classifications and theories of molecular and organic phenomena, so the pragmatic lawyers urged the need for making the law adaptable to changing social conditions in the light of modern science and logic. In Darwinian fashion, they interpreted the law as a human instrument for adjusting conflicting desires in the struggle for existence among men. Against a static conception of the law as disembodied reason or eternal natural law discoverable by the jurist and applied syllogistically to each new case, they applied the more empirical notion of man's fallible groping for order and justice in the intense competition of the market place. Justice would be no more than what intelligent compromise and social expediency could naturally yield. While the courts and other high places of government were often condemned for shady transactions sanctioned by corrupt bossism—for example, the Tweed Ring—the philosophical lawyers of the Metaphysical Club turned to "the scientific study of law," to quote the terms of the new Nathan Dane Chair set up at Harvard's Law School. The law was regarded by them as an evolving body of custom in line with the viewpoint of Sir Henry Sumner Maine's historical studies. The inductive logic of the British lawyer-philosophers, Sir Francis Bacon, Thomas Hobbes, and Jeremy Bentham, as well as their utilitarian ethics, were fused with an historical evolutionary approach to the law by our Harvard lawyers. They aimed like their British forebears to elevate the intellectual and social prestige of their calling, to confer upon it the dignity of science, though it required the most stalwart faith and practical idealism to persist in such an aim.

Our discussion of this idealistic Yankee venture of Harvard's lawyer scholars and its role in the genesis of pragmatism will begin with the problems of legal philosophy about 1870 which center about the neglected figure and writings of Nicholas St. John Green.*

Peirce, in his account of the Metaphysical Club, spoke of Green as

one of the most interested fellows, a skillful lawyer and a learned one, a disciple of Jeremy Bentham. His extraordinary power of disrobing warm and breathing truth of the draperies of long worn formulas, was what attracted attention to him everywhere. In particular, he often urged the importance of applying Bain's definition of belief, as "that upon which a man is prepared to act." From this definition, pragmatism is scarce more than a corollary; so that I am disposed to think of him as the grandfather of pragmatism.[2]

William James said, "Green went off far too young"—Green died in 1876, a year after his lifelong friend Wright. Both born in 1830, Green and Wright had known each other as schoolboys in Northampton, and had been together at Harvard as undergraduates, as University lecturers under President Eliot, and as intimate neighbors in Cambridge. We shall soon see that Holmes expressed a high opinion of Green's historical analysis of the laws of libel and slander. Holmes's close friend, John Chipman Gray,[3] who probably wrote the obituary notice of Green for the *American Law Review*, must have known of Green's and Wright's interest in Maine's *Ancient Law*,[4] Bain's psychology, Mill's logic and utilitarianism, and Darwin's evolutionary empiricism:

He [Green] was as important a figure in the field of jurisprudence as his equally lamented friend Chauncey Wright was in that of science and philosophy . . . He became a student of history, political economy, psychology, and logic, prejudice gave way to philosophy, and his convictions without losing in strength were tempered by an appreciation of the other side . . .

He handled a question of law not only with the mastery of a logician who easily reduced a case under established principles, but also, and with equal power, *in the light of the history which explains those principles,* and the considerations of political science and human nature which justify them[5] [my italics].

* Since so little is recorded of the life of "that acute and learned lawyer," as J. B. Thayer described him, I refer the reader to Appendix F for the interesting biographical facts kindly supplied me by Professor Frederick Green, the son of our subject.

When Green died in 1876 he had completed the editing and annotation of the first two volumes (1874–75) of a huge project, the annual publication of *Criminal Law Reports, Reports of Cases determined in the Federal and State Courts of the United States, and in the Courts of England, Ireland, Canada, etc. With Notes.* The *historical* approach which differentiates our nineteenth-century American jurisprudence in its evolution from the more analytical systems of the pre-Darwinian British empiricists (Bacon, Hobbes, Bentham, Austin, James Mill), is plainly expressed by Green in the preface to the first of his two volumes:

> In future volumes, no case which is deemed to be even of the slightest importance will be omitted. Doubtful points will also be commented on in notes, in which an endeavor will be made to treat the subject *historically as well as theoretically,* inasmuch as it is believed by the editor that only by the study of the history of the English Criminal Law can its excellences or its defects be clearly understood [6] [my italics].

It is important to notice that the *historical* approach is intended by our pragmatic jurist to function as an inductive aid and not as a substitute for *analysis* of legal rights and liabilities as provided in the present law. Green's pragmatic historicism does not prevent him from following Austin's logical procedure of systematizing the common law, which was "a mass of details," for, like all English law, "it has none of those large coherent principles which are a sure index to details." The "details," however, are not simply the mass of law but its shifting historical contexts with its diverse expressions of the public will. Warner, in his review of Holmes's *Common Law,*[7] warns us that the historical, evolutionary approach may prove too much, since we are always unsure whether proposed innovations will be able to overcome the dead historical weight of old lines and precedents. Finally, Holmes himself, though he did so much to breathe historical vitality into the understanding of the common law, reminds legal students that historic continuity is a necessity but not a duty. By 1880 the pendulum of evolutionary thought had swung to an extreme historicism as a result of the idealistic and theistic lucubrations of John Fiske and other evolutionary theologians. Both Green and Fiske were intimate friends of Chauncey Wright, but there is a closer affinity between Wright's and Green's analytical empiricism than between Green's pluralistic and Fiske's monistic historicism.

I shall endeavor to show from a detailed examination of Green's legal essays and notes how he kept a pragmatic balance between the analytical and historical schools of jurisprudence. In thus avoiding the lifeless formalism of the Austinians, and the metaphysical tendencies of the post-Kantians to force history into a priori schemata, Green paved the way for the sociological, empirical, and pluralistic method which was soon to be formulated by Peirce, James, and others as the consciously philosophical doctrine of pragmatism.

Green conceived that the function of philosophy in law is to give careful logical analyses of concepts and rules proceeding from slowly changing social, historical conditions. The concepts require strict definition and the rules inductive corroboration from legal precedents studied historically. *A Treatise on the Law of Negligence* by Shearman and Redfield is criticized by Green[8] for claiming to be philosophical when it is not, because it is not "analytical"; it confounds "negligence" in the *legal* sense (which is primarily concerned with liability) with the *factual* sense of neglect, and introduces into common-law cases the distinctions between degrees of negligence in civil law. Thus, it is "almost impossible to reconcile the cases in this branch of the law, even where there is in reality no conflict."[9]

There is, perhaps, no better illustration of the union in Green of historical erudition and logical analysis oriented toward a clearer understanding of the important idea of causation and responsibility in the study of law, than his article on "Proximate and Remote Cause," published in 1870.[10] This was a year before Peirce's review of Berkeley, in which he declared for "realism" of the medieval Scotist kind, and also orally presented to his "Metaphysical Club" his views on "The Fixation of Belief" and "How to Make Our Ideas Clear," originally also marked by socio-historical and analytical considerations.

In his article on "Proximate Cause and Effect" in Baldwin's *Dictionary of Philosophy and Psychology*, Peirce showed (about thirty years after Green's article of 1870) the relationship between pragmatism and this legal and obscure Aristotelian term in a way that strongly suggests the influence of Green's essay on the same theme:

> *Proximate cause and effect:* an obscure term, like most of the terms of Aristotelianism, which acquired some practical importance owing to the courts holding that a man was responsible for the proximate effects of his actions, not for their remote effects. This ought to determine what should be meant by *proximate cause and effect;* namely, that that which a man ought to have foreseen might result from his action is its *proximate effect.* The idea of making the payment of considerable damages dependent upon a term of Aristotelian logic or metaphysics is most shocking to any student of those subjects, and well illustrates the value of PRAGMATISM.

Green's historical research into the origin of the legal maxim which regards the proximate but not the remote cause as pertinent to the law ("In jure non remota causa, sed proxima, spectatur"), showed that that maxim does not appear to have been used in the English law prior to Bacon's time.[11] The maxim itself originally occurs in Bacon's *Maxims of the Law,* and illustrates what Green hails as the methodologically sound precept of Bacon "not to take a law from the rules, but to make a rule from the existing law." [12] Then Green points out that the historical question of the origin of the rule or maxim is not as important as its general application and employment by the courts as an authoritative rule in cases of maritime insurance, in actions for negligence, and in determining the defendant's liability for damages. In such cases it is often used with a different meaning, because it receives a different *application* from any intended by Bacon. What Bacon meant by it Green shows in ten closely packed pages of erudite historical references to Aristotle and his scholastic commentators.[13] Bacon not only adopted their four-fold logical and metaphysical classification of causes (material, formal, final, and efficient) but also their view of proximate cause as a *sufficient* condition for the inevitable production of an event. Green's point is that metaphysical definitions of causation are irrelevant to the purpose of the law concerned with the rights and liabilities of parties to proceedings which have nothing to do with proximity of events in space or time when viewed as mechanically connected together like the links of a chain. There is nothing in na-

ture which corresponds to this single unbreakable chain of causation. "Such an idea is a pure fabrication of the mind." [14]

The Aristotelian distinction between proximate and remote
cause retained by Bacon was analyzed by Green to mean that a
proximate cause involves (1) necessary production, (2) a connection between it and the effect which is plainly intelligible, (3)
that from which the effect can be deductively inferred; whereas,
a remote cause is one from which no necessary effect can be
clearly and conclusively drawn. It is to the a priori idea of a single necessary chain of causation which Bacon imported into his
legal maxim from scholastic philosophy that Green objects, for
he commends Bacon for seeking to avoid the misty generalities
of remote or ultimate causes. After indicating that the law does
not look to physical necessity, Green adds:

> The law in the application of this maxim [of proximate cause], is not
> concerned with philosophical or logical views of causation. When the maxim
> is applied, the whole body of facts has been ascertained by testimony. The
> facts are the subject of inquiry for a single purpose. That purpose is to
> determine the rights and liabilities of the respective parties to the proceed-
> ings. These facts alone are viewed as causes and effects which have direct
> bearing upon those rights and liabilities. . . . The difficulty is not owing
> to any great ambiguity in the meaning of the word cause. That word is
> used in its popular signification. [15]

Bacon's maxim that the law looks only to proximate causes,
says Green,

> when applied in actions of contract, is essentially a rule of construction. It
> is the same thing to say a thing comes within a contract as it is to say the
> contract embraces the thing. It is the same thing to say the loss is a proxi-
> mate consequence of a peril insured against, as it is to say that the parties
> intended that such a loss be covered by the policy. The different forms of
> expression are one in meaning . . . In actions for negligence, a defendant
> is held liable for the natural and probable consequences of his misconduct.
> In this class of actions his misconduct is called the proximate cause of those
> results which a prudent foresight might have avoided. It is called the
> remote cause of other results.
>   In determining the amount of damages in an action of contract, the
> breach of contract is called the proximate cause of such damages as may
> reasonably be supposed to have been contemplated by the parties. If there
> are other damages, of these it is called the remote cause.
>   There is no settled rule for the application of the maxim in determin-

ing the damages in actions of tort. In such actions, the damages, which are called proximate, often vary in proportion to the misconduct, recklessness or wantonness of the defendant . . . We cannot add clearness to our reasoning by talking about proximate and remote causes and effects, when we mean only the degree of certainty or uncertainty with which the connection between cause and effect might have been anticipated.[16]

Now the scholastics and Bacon differed only in the manner in which they proposed to investigate causes. The theologians used an a priori method of assuming causes, whereas Bacon laid down a system of enumeration of instances. Green sharply and pragmatically asserts:

Neither method is of practical use. It is impossible to find the art of making gold by blind experiment. It is equally impossible to find it by enumeration of instances of yellowness, of weight and of ductility. When Bacon gave it as his opinion that it was easier to make silver than to make gold, because silver was the simpler metal, he reasoned like a schoolman . . . who had not yet attained to subtlety.

There is but one view of causation that can be of practical service. What one of the various circumstances . . . we shall single out as the cause, to the neglect of other circumstances, depends upon the question for what purpose we are investigating.[17]

The purpose of research into causes in physical sciences is quite different from that in the law. In physical sciences, including human physiology, a search for proximate causes is a search for the conditions immediately antecedent to and concomitant with the effect. This is a mechanical view of causation whose logical and metaphysical core is spatio-temporal contiguity. In the law there is no search for such mechanical causes. Whatever facts of spatio-temporal character are involved are ascertained by testimony and the jury. The legally relevant facts are the subject of inquiry for a single purpose, namely, to determine the *rights and liabilities* of the respective parties involved. Green is quite positivistic and impatient with unclear metaphysical terms or unscientific, inflexible distinctions:

One difficulty is that philosophy and metaphysics are sometimes brought into a discussion to which they do not belong. Another is, that cause and effect are often viewed as parts of a "chain of causation," and the discussion thus becomes meaningless. The chief difficulty, however, is that the term proximate and the term remote have no clear, distinct, and definable significations. The division is neither scientific nor logical. Above all, it is not a fixed and constant division. It varies in different classes of actions.[18]

The meaning of the terms "proximate" and "remote" is contracted or enlarged according to the subject-matter of the inquiry. This is as pluralistic, operational, and contextual as any logical analysis of legal terms one can find in this early period of the evolution of jurisprudence in the United States. Of course, the historian of pragmatism in the law will not identify these early beginnings with the mature twentieth-century forms of pragmatic legal philosophy.

There is, then, in Green's view of causation an indefinite *plurality* of interwoven conditions rather than a single causal chain leading inevitably to any fact; there are as many different causes as there are diverse points of view and purposes in the relations of the investigator to the subject matter of his inquiry. Physical causes are sufficient only from the mechanical point of view of getting at the spatio-temporal conditions in the neighborhood of an event, but are not sufficient to decide the *legal* question of responsibility or liability under the doctrine of so-called proximate cause. Suppose a fire breaks out in a shop covered by an insurance policy which as a contract stipulates in more or less standardized clauses the rights of indemnity and liabilities of the insured shop owner and insurance company, respectively. What circumstances concerning the fire may be properly called the proximate cause in the legal sense of the loss of property insured against, is not simply a question of physical causation; for the fire might have been deliberately planned by the owner and such misconduct is a more relevant proximate cause which does not give the owner the right to collect indemnity. It would be publicly dangerous to encourage arson in that manner. The class of interests covered by an insurance policy and the particular intent of the parties are "subservient to the public bearing on the question." The main point is that law is primarily a matter of social behavior rather than physical causation: "Law is made for the protection of society," hence a physician treats a patient with a different purpose from that of a judge who seeks to prevent future wrong to society.[19]

What is the source of all our knowledge of any kind of causes? Green's answer is unequivocally based on an empirical statistical theory which underlies practical judgment, with full recognition of its tentative and fallible character. It goes to the heart of the pragmatic theory of the essential contingency of empirical truth, so prominent in Wright, Peirce, James, and Holmes.

Our anticipations of the future are founded upon our experience of the past. The experience of the past is the experience of the successions of causes and effects which always surround us. We can estimate, with a reasonable degree of certainty, the probabilities of the future occurrence of many of these successions. About successions of this kind men make contracts. Large classes of such successions can be grouped together, and the order and frequency of their happening can be predicted from past experience with something which approaches to mathematical precision. With events of this kind, underwriters deal. Experience also teaches us that various effects, which we can foresee with a greater or less degree of certainty, will or may follow from our own acts. The law makes us responsible for those effects of voluntary acts which might reasonably have been foreseen, or which are of a kind analogous to effects which might have been foreseen. There is generally no other way of determining whether events analogous to them in kind, were or might have been anticipated or foreseen, than by an appeal to experience. By applying this maxim, we make that appeal . . . The use of the maxim is liable to lead to error by withdrawing the attention from the true [legal] subject of inquiry. . . . But this is an inconvenience which must be submitted to by those who attempt to make a practical application of the maxim.[20]

Green's empiricistic pluralism (a designation he himself never employed because he eschewed labels and always went straight to the objective analysis of a specific problem) is clearly illustrated in his review of Dr. I. Ray's *Treatise on the Medical Jurisprudence of Insanity* (5th ed., 1871) in Holmes's *American Law Review*.[21] Where the physician is concerned with the abnormal behavior of his deranged patient, the legislator and the judge are concerned only with the effects of such behavior on the community. Along with Bain and Maudsley, the authorities respectively on normal and abnormal psychology in the seventies, Green is willing to accept a physiological theory of the mind, provided we realize how limited such theories are with respect to drawing any absolute line between sane and insane. We go beyond science and experience if we regard the mind as a metaphysical entity deprived of all freedom or accountability when it disappears suddenly in insanity. The mind is no such entity, and volitional functions never disappear completely while there is fear of punishment. Mental action is dependent on nervous structure; sanity and insanity are but matters of degree of stability or deterioration of the nervous system. We know no more about insanity than about its dangers to the public. The legal difficulty of ascertaining with any degree of precision how much knowledge a person must have to be held responsible for his actions is the same as the scientist's difficulty of drawing a line between the sane and the insane. "All things in nature, all things external, all ideas, all sensations, all emotions, shade into each other by imperceptible degrees. No absolute line can be drawn. Things are separated from one another by a debatable ground. The difficulty is inherent in the nature of man. Human power can make but an approach to precision." [22]

This philosophical statement of Green's can best be explained as due to the prevailing evolutionary conception of the merging of species through natural selection, a conception which was spreading into psychological and social sciences, as we have seen. Green's intimate neighbor, Wright, was busy at the time defending Darwin against the scholastic apologist, Mivart, on the central thesis of the temporal continuity of species by the cumulative and gradual effects of minute variations useful for survival. Green frequently visited another neighbor and friend, Fiske, who was lecturing at Harvard on evolution in society. Then also, the archmetaphysical evolutionist Peirce, who admired Green as an acute Benthamite member of the Metaphysical Club, was telling the Club how to make our ideas clear by breaking away from the authoritarian intuitionists and infallibilists who opposed evolution.

Jeremy Bentham was the strongest protagonist of empiricism and utilitarianism in British ethical theory and legal philosophy before Darwin and Sir Henry Sumner Maine. Bentham's definition of utilitarianism was clear and concise (his very style seems to have served as a model to Green), and opposed to the doctrine of an eternal law of nature divorced from the physical good of man, that is, man's sensory, social, and intellectual pleasures, according to the antimetaphysical Bentham. What the social and pragmatic post-Darwinian thinkers did was to generalize the idea of utility so that it became a factor in both biological and social evolution. Common to Bentham and Green was a utilitarian, antiauthoritarian liberalism. Green subscribed as completely to the empirical ideas of Bentham concerning the ends of legislation as he did to the evolutionary ideas of his own time concerning the historical roots of the law. And as Bentham always applied the test of empirical, social consequences, so we find Green waiving the illusory certainties of scholastic jurisprudence.

The best the law can do is to inquire into the power of the defendant to "foresee the usual and probable consequences of his acts," [23] before acquitting or condemning him. Whenever there is a margin of doubt, the question is answered "in favor of humanity, by giving the defendant the benefit of that doubt." [24]

Green was not simply a logician, lecturer, and practitioner of the law. Underneath the rigorously clear, historically erudite, and masterfully legal competence of his writings on the laws of tort and crime, there appears the humane liberal heart and faith in the power of reason to correct the shortcomings of and introduce a more just order into that incoherent mass of historically accumulated court decisions which constitute the common law. Though he quotes Austin with favor in this connection, Green insists that we must understand also the peculiar contortions and ungainly growth of the law *historically*, "in the way we account for the distorted shape of a tree—by looking for the special circumstances under which it has grown, and the forces to which it has been exposed." [25] Green's evolutionary or historical supplementation of Austin's analytical method appears in his effort to *account* for "the crooked and wrenched form of the law of slander and libel": "Born of the Roman, and nurtured by the canon law, its distorted person evidences the violence with which it was torn from its nurses." [26] Going into the history and evolution of the English law, from its dark formless beginnings under the rule of the Druids, through the successive governments of the Romans, of the ecclesiastics of the medieval universal church, of the Norman conquerors, and into the subsequent separation of spiritual and temporal power, Green not only accounts for the evolution of the laws of slander and libel, but also believes one can thereby prove that the whole doctrine of legal malice is pure obscurantist, inquisitorial scholasticism.

> The defendant was punished *pro salute animae*, and the matter was not looked at in a legal, but in a moral point of view, to see if the speaking of the words was a sin. When courts of law took jurisdiction of defamation they seem to have applied to this *animus* of the Romans, or *malitia* of the canon law, the elaborate scholastic structure framed in the common law—and the doctrine of implied malice was introduced into the law of slander. They affirmed that malice was in all cases necessary to maintain the action, and to find a malice which did not exist they implied it. They were like men who should persist in viewing all things through smoked glass, and should light candles to enable them to see through the glass; if they should remove the glass they would save their candles and see at least as clearly as before.[27]

Now this is not simply history, but critical history with a utilitarian purpose. Holmes was early impressed by this historical analysis by Green of the laws of slander and libel, and warmly commended it in his edition of Kent's *Commentaries on American Law* (1873). "In an able article, 6 *Amer. Law Rev.* 593 [by N. St. John Green (July 1872)], the reason for some of the distinctions between slander and libel, and for holding certain words actionable *per se*, are thought to be purely historical, and are ingeniously explained." [28]

Green was a Benthamite, as Peirce has testified, and for Bentham the medieval inquisition into intent and malice was like "the pitchy darkness of the very earliest ages." [29] The legally important thing for Bentham and Green is the practical, social *consequences* of a man's defamatory language, and not some speculatively assumed *intended* or *implied* malice. The medieval moral point of view of authoritatively judged "sinful" language was repugnant to Green, and Holmes was soon to follow in Green's utilitarian fashion by invoking the external standards of the law as the true theory of trespass and negligence.[30] By "external standards" Holmes meant several things, each of which is expressed also and earlier by Green though not in any general doctrine of liability:

(1) A quasi-behavioristic method of ascertaining intention.—

Holmes: "An act is always a voluntary muscular contraction and nothing else." . . . "The law only works within the sphere of the senses." [31]

Green: "An intention of the mind is an idea; it cannot be perceived by the senses; it can only be inferred from acts." . . . "Every action consists as such of inward feelings and outward motions, the motions forming the evidence of the feelings." [32]

(2) Irrelevance of the internal phenomena of conscience.—

Holmes: "The law is wholly indifferent to the internal phenomena of conscience." "Malice aforethought does not mean a state of the defendant's mind . . . It is in truth . . . like . . . negligence . . ." [33]

Green: "The whole doctrine of legal malice is pure scholasticism, and obscures with a thick fog every thing it envelopes. In actions for malicious prosecutions alone has it a semblance of a meaning, while in cases of homicide it means no one knows what, and in actions of slander it means nothing." [34]

(3) The primacy of public policy over individual peculiarities.—

Holmes: "The standards of the law are standards of general application. The law takes no account of the infinite varieties of temperament, intellect, and education which make the internal character of a given act so different in different men. It does not attempt to see men as God sees them . . . When men live in a society, a certain average of conduct, a sacrifice of individual peculiarities going beyond a certain point, is necessary to the general welfare." [35]

Green: "The physician deals with a single individual considered as his patient, and has nothing to do with the community or interests of the community. If he has an insane patient, his whole attention is given to this particular cure without looking beyond. . . . But law is made for the protection of society. It is for the community and not for the individual; except as through the individual it may affect the community, and promote the safety of society at large. The defendant in a criminal case is not a patient whose moral and physical health is the sole object of solicitude. The object of legal punishment is to prevent not only the defendant, but all other persons, from violating the law." [36]

Green and Holmes both believed it was in the public interest not to exempt from liability all the acts of an insane person but only those he is incapable of controlling.[37]

Both Green and Holmes were favorably impressed by the writings of a lifelong friend of Henry Sumner Maine, James F. Stephen (1829–1894), whose important and widely read work, *General View of the Criminal Law of England* (1863), was cited by Green in favor of externally observed conduct as evidence of intent,[38] and was much admired for the same reason by Holmes. Professor Hall is the first to point out the influence of James Stephen's positivistic views of criminal law on Holmes. He could have included Green with Holmes in his historical observation that Holmes's theory of objective liability "was the product of a Utilitarianism which held expediency not only the sole objective of all law but also an adequate one. He was thus close kin to Bentham and Austin," and that Holmes "brought to the earlier legal positivism the full tide of modern social positivism, the biological and evolutionary versions predominating."[39]

I do not wish to identify Holmes's views with those of Green, but only to show certain common pragmatic features due historically to their common sources in British empiricism and evolutionary positivism.

The progress and humanitarianism which Green saw evolving slowly in the law appear in his eloquent comment on the statutory extension of property rights to married women.

> The law of the status of women is the last vestige of slavery. Upon their subjection it has been thought rests the basis of society; disturb that, and society crumbles into ruins. By the married woman's property acts, the first blow has been struck. The cheek of the idol has fallen to the ground; the thunder is silent, and the earth preserves its accustomed tranquillity. The huge idol will sooner or later be broken to pieces.[40]

Against the punishability of children at the age of seven, permitted by the criminal courts of Massachusetts even in his own day, Green argues learnedly but with scarcely suppressed moral anger, pointing out that even the barbaric and superstitious ancient opinions set the age of punishable infancy "capable of guilt" above that of seven years. Both Blackstone[41] and Austin[42] are severely criticized by Green, the former for his vague moral language, the latter for his overformal and specious logic. Both fail to realize that "a bench of judges trained in dry technical reasonings and a jury of artisans and shopkeepers versed wholly in the business of practical everyday life, would be as unfit a tribunal as could be selected for the purpose of judging of the processes of the infant mind." [43]

Bentham and Green belong to that humanitarian school of legal thinkers who rebelled against cruel and barbaric methods of punishing criminals, young or old, sane or insane. The ethical criterion of the British utilitarians, "the greatest happiness of the greatest number," originated with Beccaria whose main work was devoted to the condemnation of methods of torturing and trying criminals. [44] So we find Green arguing that we must qualify the application of the principle, "Ignorance of the law is no excuse," in the case of a woman condemned for casting a ballot in a state where women were not enfranchised, for there was no proof of criminal intent or *mens rea*, but only an attempt to exercise what she regarded as her democratic privilege and right under the Fourteenth Amendment. [45]

Green expresses himself often on legal semantics in a manner that occurs frequently in Mr. Justice Holmes's opinions. [46] Hobbes, "who had a marvelous faculty for stripping words of their ambiguity," is quoted by Green, [47] as having made clear that the meaning or legal effect of a pardon is not, as Coke said, to forgive a crime, but simply to remit the punishment for a crime, as an act of mercy. Another instance, which involves the evolutionary aspect of linguistic meaning, is Green's comment on the legal question whether a man may commit rape upon a woman asleep. The author of a work on criminal law cites the medieval Statutes of Westminster to the effect that the term "against her will" was used to mean the same as "without her consent," and since a woman asleep cannot be said to have offered consent, a man may be held for rape in that case. Green's comment is: "It may perhaps be that in the Statutes of Westminster the words 'without her consent,' and the words 'against her will' were used synonymously, but the meaning of a statute passed in the dark ages cannot be determined by the ordinary rules of verbal criticism as if it were a statute of the present day." [48] Now an historically obvious element of the meaning of "against her will" was the use of violence or overpowering force. If no overpowering force is used, then it is a new crime not covered by the old law to commit rape on a woman asleep. And Green adds an idea which Holmes frequently repeated: "The creation of new crimes belongs to the legislature and not to the courts." [49]

Green traced historically the right to repel attack on one's dwelling to ancient customs: "By the early laws of all civilized nations, and by the customs of these savage and semi-savage tribes whose usages have been intelligently observed, a peculiar immunity is given to the dwelling. A trespass upon the habitation is regarded as a kind of sacrilege, entitling the occupier to inflict summary vengeance." [50] The evolution of the idea that a man's house is his castle is traced by Green's historical references to early Greek and Roman laws of property in Fustel de Cou-langes (*La Cité Antique*), to Sir Henry Maine's anthropological study of *Village Communities in the East and West,* and to the place where the custom most strongly prevailed, namely among the Germanic tribes (Salic law and laws of Anglo-Saxons and Normans). Another illustration of Green's historical and philo-sophical learning is to be found in his view that the form and requisites of an indictment at common law are derived from scho-lastic logic and the use of Aristotle's categories.[51] A grand jury was required to ask and answer: *Quis? quid? ubi? quibus auxilis? cur? quo modo? quando?* The method of pleading exhibited in the Year Books resembles scholastic disputations.

Beside the union of the socio-historical and analytical meth-ods which is common to the legal thinking of Green, Warner, and Holmes, who thus mark an advance over the predominantly un-historical analyses of Bentham and Austin, there is a more dy-namic psychology in the American pragmatists than the passive sensationalism of the British empiricists. The nominalistic theory of ideas in Ockham, Hobbes, Locke, Berkeley, Hume, and Mill, made ideas merely representative copies of particular external qualities. The voluntaristic idealism or scholastic realism of Duns Scotus as interpreted by Abbot, Green, and Peirce, implied that ideas had a more general and dynamic power of influencing ac-tion. In the light of Green's high admiration for Scotus—which we find also in Bain's Scottish realism and in Peirce and Abbot—

it is significant that Green says: "There is always a tendency to act out an idea." [52] This tendency to act out an idea is capable of being restrained, but only by a power of control acquired by habit. When a man mentally recalls a scene which once excited his indignation, he again becomes angry, truthfully saying he is "indignant at the idea." "The infection of certain ways of committing crime and suicide is a matter of common remark." [53]

We shall find Warner saying at the conclusion of his address to the American Bar Association on "The Responsibilities of the Lawyer" [54] that ethical ideals can be realized only if "we go down among men and work." Similarly, Holmes in discussing the suppression of radical ideas because they are incitements to action which constitute a present danger, asserts that "all ideas are incitements to action," and the only relevant issue is that of "clear and present danger."

Green's definition of "a reasonable doubt" is similar to Peirce's definition of belief as the settlement of the irritation or uneasiness of doubt: "A reasonable doubt is the doubt which a juryman feels when he is not satisfied as to the truth of the fact to be established by evidence on the part of the prosecution. If he feels uneasy as to the truth of the fact in saying guilty, he has a reasonable doubt." [55]

Legal rules or maxims should not be accepted as absolutely universal. Green quotes with favor one of the authors of the *Port Royal Logic*: "Maxims are to be distrusted, for there are few general truths; all have their exceptions and their limits, and very false applications of them may be made, because the mind being taken up with the verity of the maxim, examines with little care the subject to which it is applied." [56] This caution against adhering formally to general rules is applied by Green to the legal maxim: "Every person is presumed to intend the necessary consequences of his own acts." The age and state of mind of the defendant and unforeseeableness of some consequences, qualify this maxim.

Enough examples have now been given of Green's pragmatic, pluralistic, and empirical temper of mind to show its affinity with that of Holmes. Evidence of their personal association is given not only by Peirce's reference to their membership in his "Metaphysical Club," [57] where they shared the influence of Chauncey Wright, but also by an account given by Brooks Adams of their mutual friend, Melville M. Bigelow, Dean of the Boston Law School:

> I remember Mr. Bigelow first in the old library, in the old court-house [of Boston], which was at that particular moment [Autumn of 1872] chiefly occupied by three remarkable men, by no means the least of whom was Bigelow himself. One of these was Oliver Wendell Holmes, who was then editing Kent; another was Nicholas Saint John Green, who was then lecturing in the Harvard Law School on Torts, and, by the way of parenthesis, I may say that than Green, I never listened to a greater lecturer. . . .
> Mr. Bigelow was great on the subject of the growth, or the evolution of a theory, or of an idea.[58]

Torts was given for the first time as a separate branch of the law when Green lectured at the Harvard Law School (1870–1873). He prepared an abridged edition of Addison on Torts with notes for use as a text. This called forth the famous statement by Holmes in the *American Law Review:* "We are inclined to think that Torts is not the proper subject for a law book." Evidently there was too much a priori metaphysical ethics in the theory of criminal law and torts to suit Holmes's positivistic mind at the time. Holmes changed his mind a few years later in his Lowell Lectures on the common law and did treat torts as a proper subject for a law book. Thus Green, along with Bigelow, Wright and J. Stephen, seems to have had a share in Holmes's adoption of the external standard of public morality accepted at a given time as against the intuitive and metaphysical ethics of the Scottish and German schools. Wright, in the *Nation,*[59] had attacked McCosh's intuitionism and the a priorism of German Darwinism. Moreover, Wright like Green checked the excessive claims of the analytical and historical schools of jurisprudence while doing justice to the contributions of both. Wright had pointed out that Austin and Maine were dealing with distinct sorts of problems in the law, the former with its logical coherency and ideal sanctions and the latter with its historical development out of half-instinctive, half-traditional interests, that is, out of the primary morals which are not the products of legislation. When Maine in his *Early History of Institutions* (1875) complains that Austin does not treat "sovereignty" historically, he falls into the error, Wright properly indicates,[60] "which at the present time is common enough, of confounding the province of history and of logic." Now this view of Wright's—which is the same as Green's—that such ethical principles as the "greatest good for the greatest number" are practical *ideals* to guide the legislator but are not intended to serve as explanations of the historical growth of laws and morals, is precisely what Justice Holmes expresses in his praise of Morris R. Cohen's "excellent article, 'History versus Value'": "For the last thirty years [1885–1915] we have been preoccupied with the embryology of legal ideas; and explanations, which, when I was in college, meant a reference to final causes, later came to mean tracing origin and growth. But fashion is as potent in the intellectual world as elsewhere, and there are signs of an inevitable reaction." [61]

The reaction, however, that Holmes favored was not in the direction of Cohen's semiplatonic conception of truth and justice, but in a return to the more dynamic dualism of Wright's and Green's empirical discrimination of ideals from historical facts. Logical stages of evolution permeated by final causes oversimplify the embryology of the law. "We want reasons more than life history . . . By the time a proposition becomes generally articulate it ceases to be true—because things change about as fast as they are realized." [62] Since men generally believe what they want to believe, absolute ideals of conduct and the absolute truth neither justify nor explain the making of laws which adapt men to their conflicting and changing needs. Green's legal reflections, attempting to do justice to both the logical and evolutionary aspects of the law as an instrument in the service of social needs, are in line with the kind of functional thinking or *logica utens* which was being formulated in more general, philosophical terms by Peirce in logic, by James in psychology, and by Holmes in the law.

CHAPTER EIGHT

Evolutionary Pragmatism in Holmes's Theory of the Law

Unless we pedantically and unhistorically reserve the ancient name "philosopher" for system-builders or academic professors of philosophy, Oliver Wendell Holmes, Jr., is certainly a philosopher in his own right. Many a tribute has been paid to his philosophical addresses and writings by those who knew and admired him as a lawyer and judge deeply concerned about the larger ramifications of his calling. His correspondence with professional philosophers, William James, Max C. Otto, and Morris R. Cohen, as well as with legal scholars, Sir Frederick Pollock, Harold J. Laski, and others, leaves no doubt that he sought and formulated for himself answers to the perennial questions of philosophy: What is truth? What is man's place in the cosmos? What is the *summum bonum*, and how is it related to the evolution of law? Is there any evidence of purpose in evolution which might be "squeaked" out if we "twisted the tail of the cosmos" with a sufficiently powerful metaphysical pull?

Holmes, like James, abhorred traditional systems of static law and absolutistic metaphysics. New England's Platonism he learned early to view with skepticism; witness, his youthful critical essay on Plato's *Republic*[1] which he showed to Emerson and which provoked the seer's comment: "When you strike at a King, you must kill him." The great progress of mid nineteenth-century science was at the bottom of Holmes's skepticism of metaphysical absolutism. He did not have to read Helmholtz on the conservation of energy, Darwin on the origin of species, Spencer on cosmic evolution—he tells us he did read Buckle— for the ideas contained in these epoch-making works were "in the air" during Holmes's college days. Discussion of these important scientific ideas was frequent in his talks with his British friends, Leslie and Fitzjames Stephen, and Frederick Pollock, as well as with his brilliant Harvard companions, including the members of Peirce's Metaphysical Club and other clubs in Boston to which he, William James, and John Chipman Gray belonged.

Morris R. Cohen, in one of the many letters that passed between him and Holmes from 1915 to 1934, inquired (in 1919) whether the reading of Voltaire had had any direct influence in producing Holmes's skepticism. Holmes's reply was:

> Oh no—it was not Voltaire—it was the influence of the scientific way of looking at the world—that made the change to which I referred. My father was brought up scientifically—i.e. he studied medicine in France— and I was not. Yet there was with him as with the rest of his generation a certain softness of attitude toward the interstitial miracle—the phenomenon without phenomenal antecedents, that I did not feel. The difference was in the air, although perhaps only the few of my time felt it. The Origin of Species I think came out while I was in college—H. Spencer had announced his intention to put the universe into our pockets—I hadn't read either of them to be sure, but as I say it was in the air. I did read Buckle— now almost forgotten—but making a noise in his day, but I could refer to no book as the specific cause—I have never read Voltaire and probably at that time had read nothing. Emerson and Ruskin were the men that set me on fire. Probably a sceptical temperament that I got from my mother had something to do with my way of thinking. Then I was in with the abolitionists, some or many of whom were sceptics as well as dogmatists. But I think science was at the bottom. Of course my father was by no means orthodox, but like other even lax Unitarians, there were questions that he didn't like to have asked—and he always spoke of keeping his mind open on matters like spiritualism or whether Bacon wrote Shakespeare—so that when I wanted to be disagreeable I told him that he straddled, in order to be able to say, whatever might be accepted, well I always have recognized, etc. which was not just on his part.[2]

The clash between father and son may be regarded as symbolic of the impact on New England's transcendentalism of the positivism encouraged by the new theories of physics and biology. The older Holmes at times argued with poetic eloquence against the extension of the mechanical conception (which he accepted fully for the material world of science) to the realm of moral freedom and spirit. The autocrat of the breakfast table taught his medical students that mechanical force—the *vis viva* of Descartes, Leibnitz, and Helmholtz—"is only the name for the incomprehensible cause of certain changes known to our consciousness, and assumed to be outside of it. For me it is the Deity himself in action." [3] Young Holmes regarded *vis viva* or the conservation of energy as in no need of divine concurrence. Even during his Civil War campaign he found time to taunt his father with a mathematical puzzle that would prove $1 + 1 = 3$: two Boscovitchian points of force generate a third at their center of gravity.[4] In his diaries of 1866 and 1867 while traveling in Europe, and in his letters to William James, Holmes often refers to *vis viva*.[5] The great progress of the physical and social sciences in Holmes's youth, including the works of Darwin, Buckle, Comte, Maine, Bain, Mill, *et aliorum*, prepared the way for his predictive theory of law.

While the elder Holmes and Asa Gray did defend Darwin against Agassiz at the Saturday Club, they refused to admit philosophically any conflict between Darwin and their own Unitarian faith in the ultimate harmony of science and religion. Not so with young Holmes, who seems to have followed the more skeptical positivism of Chauncey Wright's defense of Darwin. About fifty years after the death of Wright (1875), we find Holmes on three occasions referring to Wright with full approval: "Chauncey Wright, a nearly forgotten philosopher of real merit, taught me when young that I must not say *necessary* about the universe, that we don't know whether anything is necessary or not. I believe that we can *bet* on the behavior of the universe in its contact with us. So describe myself as a *bet*-abilitarian." [6]

This skeptical antimetaphysical view of necessity which Holmes learnt from Wright safeguarded him from espousing the theistic and cosmic evolutionism of Fiske and Peirce. An evolutionary and pragmatic empiricism furnished the philosophical substance of Holmes's theory of history and law which makes a sharp break with German dialectical formalism—"a syllogism can't wag its tail"—and the classical doctrine of natural law. The linkage of Holmes with the Metaphysical Club lies in the fact that Wright's arguments for the ethical neutrality of science and evolution appear in scarcely modified form in Holmes's separation of ethical ideals from the science of law, which uses only external standards of social expediency and the public force. Both Wright and Holmes rejected Peirce's and Fiske's evolutionary theology and James's will to believe as sophisticated forms of wishful thinking.

It was a natural thing for Holmes to admire the positivistic ideas of Wright and Green. Both were emancipated from theology and a priori metaphysics in their discussions of the fundamental problems and methods of the natural and social sciences, and Holmes saw in their secular naturalism and utilitarian ethics the path of enlightenment in the growth of the law. With Green and Wright, Holmes accepted the utilitarianism of Bentham and Mill alongside of the nineteenth-century evolutionary historicism of E. B. Tylor and Sir Henry Maine. Holmes took the analytic method of pre-Darwinian English positivistic thinkers and the comparative genetic method of evolution as furnishing the twin keys to law as an evolving institution and as an anthropological document for the science of jurisprudence.

Sir Henry Maine regarded competition as "the regulated private war of ancient society broken up into undistinguishable atoms." [7] The study of the customs of primitive man was far from revealing anything like the brotherhood or equality of men preached by reformers. E. B. Tylor had concluded his two-volume study of *Primitive Culture* (1871) with an admonition that anthropology was essentially "a reformer's science." There was so much to clear away in social institutions of the accumulated "survivals" of ancient custom, ritual, myth, and superstition, residues of evolutionary stages in the development of civilization. Sir Henry Maine had argued a decade earlier that the investigation into the origin and growth of social and juridical institutions and conceptions would throw a great deal of light on the present social mechanism. And as one important consequence of the new comparative method of anthropological inquiry, Maine believed that "some celebrated maxims of public policy and private conduct, which contain at most a portion of truth, might be revised and corrected." When Maine cited as an illustration of such maxims the Benthamite principle of the greatest happiness for the greatest number, Wright came out against any application of evolutionary method that might confuse the role of historical theory with that of practical ethics. Wright remarked of Maine's work that "a stronger plea for the value of abstract studies in a practical community like our own, trying to work out salvation by democracy and equality, could hardly be made." [8]

Holmes also showed the same naturalistic, evolutionary attitude to the relation of logic to law as appears in the following

reflections of Wright concerning the wisdom of legislating eu-
genic marriages:

> Reason, or rather the analytic reason, is but a rude guide of life; there
> can be but little *positive* wisdom in any system of laws; and the happiness
> of life in minutiae depends upon very much that is not common in our
> judgments, or laid down in any defined wisdom; these are not competent
> to deal with such questions . . .
> No lawgivers, no private counsellors, are at all equal to the subtle
> skill of nature, shown in the survival of the fittest; which, though a rough
> remedy for evils that wisdom, if it existed, might forestall, is one which
> wisdom has not yet equalled. The ancient state of Sparta, whose lawgivers
> undertook to do the work of nature in selection, perished in consequence.[9]

Holmes shared with Wright an Emersonian conception of
a wisdom in nature's evolutionary processes, even when they appear to be most ruthless in eliminating the unfit. Holmes, and
Wright also, are skeptical of the power of analytic reason to overcome the strength of the unreasoning mass of the public's will and
emotions. In so far as the law embodies the historical expression
of such will and impulses as the public force imposes on those
who violate the law, "the life of the law is not logic but experience";[10] namely, the massive, accumulated historical experience
of men who carry on the evolutionary process in the competition
of the market place. Wright's view of the economic side of the
class struggle for survival in society probably influenced Holmes,
who lived through sixty years more of the growing industrial
strife of our modern United States. For we have seen that Wright
was not alarmed about the revolution of the Paris Commune. He
defended the right of private property to legal protection so long
as it performed the publicly useful function of production, and he
condemned its abuse by unproductive and prejudiced aristocracies who resort to "legalized robbery." Holmes regularly argued against adopting a socialistic program which ignored the
cost of a revolutionary overthrow of the system of private property. Capitalism for Holmes did in its evolution increase the sum
of consumer's goods, for the figures showed a greater public
distribution of these goods. Holmes went so far, however, as to
say that if the public mind ever evolved to the point where it
wanted socialism, he would vote and fight legally against it, but
would accept it only if the public force prevailed in a majority

show of strength. The "social Darwinism" of Holmes appeared in his Malthusian critique of socialism: "I believe that Malthus was right in his fundamental notion, and that is as far as we have got or are likely to get in my day. Every society is founded on the death of men . . . I shall think socialism begins to be entitled to serious treatment when and not before it takes life in hand and prevents the continuance of the unfit." [11] He certainly supported the growing control of big business by government in the Lochner case: "The Fourteenth Amendment does not enact the principles of Spencer's *Social Statics*." [12] Twentieth-century liberals like Justice Frankfurter saw in Holmes's early dissenting opinions the historical germs of the laws that grew into some of the later social legislation of the New Deal.

Wright had cautiously issued a rationally pragmatic warning to remember the Delphic oracle to those whose lives are predominantly practical and executive: "Know first, and act only on real knowledge; beware of opinion." There is the same skeptical cautiousness in Holmes, who regarded the law as a prophecy of the way the public force, as expressed by the courts, would exert itself against those who wished to change prevailing custom as well as against those who resisted changes demanded by public need. There seems to be a paradox in Holmes's enlightened skepticism toward social policy, for it leads to conservatism when the changes are resisted, and to radicalism when the public's changing demands override established policy. Holmes's predictive theory of the law as "what the courts will do in fact" does not resolve the paradox, because it is neutral, as a logical principle should be, with respect to what sort of policy the legislature and the courts should want to have enforced. On the whole, however, Holmes inherited that form of social evolutionism which favored gradual rather than revolutionary change. Change for the better is best attained by free trade in ideas. Though the social test of truth is the evolutionary power of thought to get itself accepted in the competition of the market, it is still truth rather than blind willful seeking after power which furnishes the only safe ground upon which men's wishes can in the long run be carried out. That, at any rate, was Holmes's theory of our Constitution: "It is an experiment as all life is an experiment." [13] Hence its interpretation is subject to continual revision in diverse historical contexts.

Holmes's classical sentence, "The life of the law has not been logic: it has been experience" is not a repudiation of reason, but a historical statement best understood in the evolutionary context of the 1870's. What Holmes meant by "logic" is clearly stated in the two sentences preceding the one just quoted from the first paragraph of his great work, *The Common Law;* logic is one of the tools needed to obtain a general view of the common law: "To accomplish the task, other tools are needed besides logic. It is something to show that the consistency of a system requires a particular result, but it is not all." Now every student of logic knows that a true empirical generalization requires more than logical consistency, though it cannot dispense with it. What Holmes wished to drive home was not this elementary logical truism, that empirical premises are required for empirical generalizations, but the kind of "experience" that makes up the substance of the law: "The felt necessities of the time, the prevalent moral and political theories, intuitions of public policy, avowed or unconscious, even the prejudices which judges share with their fellow-men, have had a good deal more to do than the syllogism in determining the rules by which men should be governed." The evolutionary and pragmatic empiricism of Holmes appears clearly in the rest of the paragraph:

> The law embodies the story of a nation's development through many centuries, and it cannot be dealt with as if it contained only the axioms and corollaries of a book of mathematics. In order to know what it is, we must know what it has been, and what it tends to become. We must alternately consult history and existing theories of legislation. But the most difficult labor will be to understand the combination of the two into new products at every stage. The substance of the law at any given time pretty nearly corresponds, so far as it goes, with what is then understood to be convenient; but its form and machinery, and the degree to which it is able to work out desired results, depend very much upon its past.[14]

Holmes's evolutionary empiricism is pragmatic because, like Wright and Green, he regarded the history and theory of the law as instrumental in understanding and revising it as an evolving institution: "I shall use the history of our law so far as it is necessary to explain a conception or to interpret a rule, but no further . . . The study . . . is necessary both for the knowledge and for the revision of the law." [15]

A leading illustration of his evolutionary approach to the common law appears in his theory of early forms of civil and criminal liability, the first of the dozen Lowell Lectures that went into the writing of his major opus. In that first chapter Holmes shows, after a study of the old forms of legal procedure in the Roman law and German customs and their English offspring, that the main ground of the law was the primitive desire of vengeance or retaliation against an offending thing, animate or inanimate. An ox that injures a person was stoned or surrendered to the injured or his relatives, and a similar fate awaited a slave; under the law of the Twelve Tables, an insolvent debtor could be cut up and his body divided among his creditors.[16]

It will readily be imagined that such a system as has been described could not last when civilization had advanced to any considerable height . . . Masters became personally liable for certain wrongs committed by their slaves with their knowledge, where previously they were only bound to surrender the slave . . . Still later, ship-owners and innkeepers were made liable *as if* they were wrong-doers, for wrongs committed by those in their employ on board ship or in the tavern, although of course committed without their knowledge.[17]

The evolution of the laws of liability is explained by

considerations of what is expedient for the community concerned. Every important principle which is developed by litigation is in fact and at bottom the result of more or less definitely understood views of public policy; most generally, to be sure, under our practice and traditions, the unconscious result of instinctive preferences and inarticulate convictions, but none the less traceable to views of public policy in the last analysis.[18]

Ignoring the evolutionary and social function of the law, "the official theory is that each new decision follows syllogistically from existing precedents. But just as the clavicle in the cat only tells us of the existence of some earlier creature to which a collarbone was useful, precedents survive in the law long after the use they once served is at an end and the reason for them has been forgotten." [19]

In Holmes's realistic and evolutionary conception of the common law there is admittedly a brute element of force in the external standards of the law. Force and externality enter into the

struggle for existence. Peirce put them in his category of Secondness, but Holmes had little use for Peirce's metaphysical categories. The use of force by the law is, like any act, neither good
nor bad per se,[20] but becomes so in an external relation to public
needs and welfare: "The police power extends to all the great
public needs. It may be put forth in aid of what is sanctioned by
usage, or held by the *prevailing morality or the strong and preponderant opinion to be greatly and immediately necessary* to
the public welfare." [21]

In his account of the criminal law, Holmes starts with the
Benthamite recognition of the desire for vengeance, expressed
by Sir James Stephen: "The criminal law stands to the passion
for revenge in much the same relation as marriage to the sexual
appetite." [22] What Holmes proceeds to show is that the law has
evolved from personal desire for vengeance to the more objective
need of public safety. The external standard of the law looks only
to the consequences of an act hurtful to society rather than to a
man's personal morality or to the state of his consciousness on the
matter. Under the external standard of the law a man acts always
at his peril if he fails to conform to what is generally expected of
the average prudent person.

> The standards of the law are standards of general application. The
> law takes no account of the infinite varieties of temperament, intellect, and
> education which make the internal character of a given act so different in
> different men. It does not attempt to see men as God sees them, for more
> than one sufficient reason. In the first place, the impossibility of nicely
> measuring a man's powers and limitations is far clearer than that of ascer-
> taining his knowledge of law, which has been thought to account for what
> is called the presumption that every man knows the law. But a more satis-
> factory explanation is, that, when men live in society, a certain average of
> conduct, a sacrifice of individual peculiarities going beyond a certain point,
> is necessary to the general welfare.[23]

Even intention to deceive or defraud must be proved in
court to have induced action on the faith of the representation,
so that Holmes does not differ from Bentham in using the objective test of tangible public consequences in both the civil and
criminal law.[24]

> The tendency of a given act to cause harm under given circumstances
> must be determined by experience. And experience either at first hand or

through the voice of the jury is continually working out concrete rules, which in form are still more external and still more remote from a reference to the moral condition of the defendant, than even the test of the prudent man which makes the first stage of the division between law and morals. It does this in the domain of wrongs described as intentional, as systematically as in those styled unintentional or negligent.[25]

Holmes seems to hedge when he adds: "But while the law is thus continually adding to its specific rules, it does not adopt the coarse and impolitic principle that a man acts always at his peril." [26] The point is that judges and juries consider it relevant to know whether the defendant had a fair chance of avoiding the infliction of harm before he becomes answerable for the consequences. Now "a fair chance" and "answerability" are matters pragmatically judged by average standards and by evolved public opinion aiming to preserve society.

The competent reviewer of Holmes's book for the *Albany Law Journal* noted that Holmes

often utilizes his extensive knowledge of the writings of the German jurists . . . In early forms of liability, the author's account of the development of legal liability at common law is given. The method of this essay is mainly historical, but the history is not confined to the conception of the common law only; it is concerned with the various archaic systems which have ultimately blended with the common law. In the succeeding lectures on the Criminal Law and on Torts, the author's method is no longer that of the legal historian, but that of the analytical and philosophical jurist. We do not know in which character to prefer the author; both are excellent in their way.[27]

Thus Holmes early showed the same pragmatic balance between the historical and analytical approaches to human evolution which we have already found to be characteristic of Wright's, Green's, and Warner's* thinking. Holmes seems to have revealed this best in his lectures on contract,[28] in which he traced the historical development of the laws of contract in order to show that some features formerly considered essential to a perfect contract have since become nonessentials; for example, the verbal ceremonial *Spondes ? Spondeo, etc.* of the Roman law. Such formalities served the practical purpose of identifying and solemnizing agree-

---

* See Appendix G for Warner's review of Holmes's method in *The Common Law.*

ments that incur legal obligations, but we have other ways (notaries' seal) for performing the same function today. The progress of "our scientific jurists," our legal reviewer concludes, is exemplified by the evolutionary approach of Holmes:

> In the chapter on contract, Professor Holmes has continued a series of scientific observations which were first brought forward by Sir Henry Maine, and subsequently applied definitely to the common law of contract by Mr. Frederick Pollock and other writers. The lectures by Professor Holmes are the complement of the works of the authors named.[29]

Evidently what made Holmes's work "scientific" was his evolutionary approach to the law and his positivistic separation of the law from ethics and metaphysics. Holmes approved Dr. Wu's idea that "Jurisprudence deals with the psychology of the average" rather than with transcendental norms.

In accordance with his bet-abilitarianism, Holmes regarded the essential feature of a contract to be the assumption of a risk by a promisor that a certain future event will come about, involving a promisee who has a right to sue for damages if the event does not come about. Holmes's theory is to be contrasted with the German metaphysical theories of contract (in Kant, Hegel, Jhering) as a meeting of transcendental wills endowed mystically with inherent rights and duties relating to ethical reason:

> Nowhere is the confusion between legal and moral ideas more manifest than in the law of contract. Among other things, here again the so-called primary rights and duties are invested with a mystic significance beyond what can be assigned and explained. The duty to keep a contract at common law means a prediction that you must pay damages if you do not keep it—and nothing else. If you commit a tort, you are liable to pay a compensating sum. If you commit a contract, you are liable to pay a compensatory sum unless the promised event come to pass, and that is all the difference. But such a mode of looking at the matter stinks in the nostrils of those who think it advantageous to get as much ethics into the law as they can.[30]

In this same essay Holmes stated most clearly his definition of the law:

> The prophecies of what the courts will do in fact, and nothing more pretentious, are what I mean by the law . . . The object of our study, then,

is prediction, the prediction of the incidence of the public force through the instrumentality of the courts. The means of the study are a body of reports, of treatises, and of statutes, in this country and in England, extending back for six hundred years, and now increasing annually by hundreds. In these sibylline leaves are gathered the scattered prophecies of the past upon the cases in which the axe will fall. These are what properly have been called the oracles of the law. Far the most important and pretty nearly the whole meaning of every new effort of legal thought is to make these prophecies more precise, and to generalize them into a thoroughly connected system.[31]

Holmes was not an ethical nihilist. He simply distinguished between what the law ought to be and what it is. He wished to be realistic about the difference between private moral preferences and the more powerful force of public opinion and custom embodied in the law. When, for example, the nation is in danger of attack by another nation and congress votes military conscription, "we march up a conscript with bayonets behind to die for a cause he doesn't believe in. And I feel no scruples about it. Our morality seems to me only a check on the ultimate domination of force, just as our politeness is a check on the impulse of every pig to put his feet in the trough."[32]

The aesthetic standards of personal morality—Holmes certainly bristled with these—often conflict with the larger evolutionary movement of the law governed by the external standard of "the circumstances in which the public force will be brought to bear upon a man through the Courts,"[33] for "no society has ever admitted that it could not sacrifice individual welfare to its own existence."[34] For Holmes, there is no Hegelian embodiment of divine reason in the state or law. "The common law is not a brooding omnipresence in the sky . . . It always is the law of some state."[35]

The law, as an evolving organic institution, disclosing in its history "every painful step and world-shaking contest by which mankind has fought and worked its way from savage isolation to organic social life,"[36] must turn as a calling of thinkers to scientific data and methods. These alone can establish questions of fact or the relative worth of conflicting claims by calculating the cost and effects of their realizations. The practitioner of law needs to know no more of the history of law than he can utilize; for example, the American lawyer will not find the Roman law as relevant to the existing legal system in the United States as the continental lawyer will. However, for Holmes, this practical consideration does not impugn the value of the scholarly investigation of the history of law for its own interest as an anthropological study.

Since "the first requirement of a sound body of law is that it should correspond with the actual feelings and demands of the community, whether right or wrong," [37] the education of the public and the introduction of new customs molded by that education are prerequisite for the reform of the law. That may be one of the reasons why Montesquieu's *Spirit of Laws* was a great favorite classic of Holmes.

The practical teachings of experience tend to settle down into rules of law, regardless of moral predilections or the prevalence of a certain ideal or theory of government or society. "Law, being a practical thing, must found itself on actual forces." [38] For instance, man like the domestic dog instinctively will not allow himself to be dispossessed of what he holds. [39] But this individualistic propensity runs afoul of the tendency of public policy to ignore the individual's private desires. The law through its external standards of what society expects of an individual's conduct does limit the freedom of individuals. A practical joker cannot yell "Fire," in a crowded theater and claim the protection of the constitutional right of free speech. And in war time, we must curtail some of the so-called inalienable rights of man: Debs was not to be condoned for interfering with military recruiting; though in peacetime pacifists have a right to be heard. It is the clear and present danger to the public interest that determines or should determine the limits of civil liberties.

It is not within my purpose or competence to add to the large literature on Holmes's legal opinions. I have tried only to exhume some of the ideas of the philosopher "touched with fire" in order to discern what evolutionism and pragmatism meant in the life of the law as he viewed it. Though no consistent system of ideas can be properly imputed to Holmes in his delightfully worded aphorisms about truth, the cosmos, or social ethics, there is plainly discernible in him an empiricism that is grounded in man's biological struggle for existence as well as in historical warfare and competition. [40] There is in him a realism that advocated the thinking of things and not mere words. There is an enlightened skepticism which led him to distrust the will-to-believe doctrine of his more tender-minded college companion, William James.

> I regard the will to believe as of a piece with the insistence on the discontinuity of the universe which Bill James shares with Cardinal Newman and which I suspect as induced by the wish to leave room for the interstitial miracle. When we were in our 20's, W. James said to me (in substance) that spiritualism was the last chance to spiritualize or idealize the world. I then and ever since have regarded that as a carnal and superficial view. As to the will to believe why may we not ask on what ground it is recommended except some assumed can't help to which we all must yield—Otherwise why would it not be a sufficient answer to say I don't want to? [41]

There is a tough-minded faith in the dependability of an ethically neutral scientific approach to social questions which accompanies his liberal hope for reform by the free trade of ideas. There is the poet of the class of 1861 who warns law students that the law is a calling for thinkers, not for poets or artists. There is the historian of the common law who tells these students that historical continuity with the past is not a duty but only a necessity. Finally, there is still in Holmes the Puritan sense of our ultimate dependency on a transcendental power of which we are the instrument and which, above all our frustrated aspirations, urges us to go down among men and work.

It was the influence of the progress of science amid the grim realities of the Civil War that turned Holmes away from the evolutionary theism of Peirce and Fiske and from the will-to-believe spiritualism of James. His predictive theory of the law and its amoral external standards became incorporated in John Dewey's more general theory of value judgments as hypotheses that require testing by their social consequences.[42] Holmes's pragmatism is less Hegelian than Dewey's, and on questions of ultimate ethical ends, less instrumental. Neither Dewey's emphasis on technology nor James's moralism appealed to the rugged Holmes of simple taste, content with the barberry bushes and granite rocks of his New England youth. "I do not believe that the justification of science and philosophy is to be found in improved machinery and good conduct. Science and philosophy are themselves necessaries of life. By producing them civilization sufficiently accounts for itself, if it were not absurd to call the inevitable to account." [43] Holmes said this in a speech to the Yale alumni in 1891, the year after James's *Principles of Psychology* had appeared with its instrumentalist theory of reasoning. He quoted it again in 1920 in his correspondence with Professor M. R. Cohen. Professor Cohen in replying also objected to James's psychologism, but took the line of Peirce's "logical pragmatism":

> My agreement with pragmatism extends to the main point made by Peirce, viz. that the way to make our ideas clear is to examine their *possible* consequences, or in technical language, all their possible implications. It is an attempt to extend the experimental method to the handling of ideas, and very fruitful if used logically, for the essence of intellectual liberality consists in the realization that what is familiar to us is only one of a number of possibilities. Logical pragmatism as a method of exploring the field of logical possibilities is, therefore, of the highest value. This aspect, however, has not been developed by James or Dewey because they are not interested in logic and metaphysics but only in psychology. [44]

This is a rather harsh judgment of James and Dewey, overlooking their liberalism because they did not take to Cohen's own semi-Platonic view of logical possibilities. Neither did Holmes, who is closer to Dewey than would appear from Cohen's sharp separation of the evolutionary and logical aspects of thought. Confirmation of the kinship between Holmes's and Dewey's philosophical attitudes is found in the letters of Holmes to his young confidant and admirer, Dr. Wu, student of law and philosophy, who asked Holmes whether Dewey's *Experience and Nature* did not express Holmes's way of thinking; Holmes replied:

> Pursuant to your recommendation I sent for Dewey's *Experience and Nature* and am reading it. It makes on me an impression like Walt Whitman, of being symphonic, and of having more life and experience in his head than most writers, philosophers or others. He writes badly and creates more difficulty by his style than by his thought. I could not give a synopsis of what I have read, yet I have felt agreement and delight even when I got only an impression that I could not express. I agree with you that he is a big fellow and I expect to believe when I have finished it as I do now that the book is a great book. [45]

More than a year later, Holmes confessed to Wu that he had read Dewey's book twice: "I thought it great. It seemed to me to *feel* the universe more inwardly and profoundly than any book I know, at least any book in philosophy." And six months after, Holmes repeated to Wu that Dewey "has the same feeling I have about the universe." [46]

Dewey's logical theory is in complete agreement with Holmes's view that "thinking is an instrument of adjustment to the conditions of life—but it becomes an end in itself." [47] In other words, thought evolves from being an instrument for biological survival to becoming the flower and fruit of a civilization. This Aristotelian theory of thought was developed by Santayana in his *Life of Reason,* of which Holmes said: "His [Santayana's] skepticisms seem all right, his dogmatisms comic—his total not quite charming and yet nearer to my way of thinking (I guess) than either of his former associates [James and Royce]." [48]

Thought is neither merely an instrument nor merely a terminal act. It is in relation to evolutionary processes that it serves as an instrument, but with the increase of leisure and freedom, thinking acquires satisfactions of its own in art, science, and philosophy. Holmes, master of similes, suggests picturing thought as a strawberry plant sending out shoots first to preserve itself, then to sprout into plants of its own. The important condition is freedom to grow and reach out for the air and sunshine. So thinking as a practical means of living and as a source of intrinsic satisfactions requires freedom of inquiry to reach out for the truth which alone can make men free.

Still, it cannot be said that Holmes's evolutionary and pragmatic conception of truth was consistently expressed. He oscillates, with equally brilliant eloquence at both extremes, between an individualist, subjective theory and a social, objective one. On the one hand, Holmes says, "My ultimate test of truth . . . is that *I can't help* believing so and so. I have learned by experience that this test is not infallible." [49] On the other hand, truth is what gets accepted in the competition of ideas in the social arena. In the former, subjective view, necessity is not completely abolished; it is simply transferred to the inner world of felt need. Now why should one's present feeling of what can't be helped be taken as *ipso facto* true or necessary? Self-deception is the commonest of human foibles. In science, and Holmes accepts science as the most trustworthy means of getting at truth in law as well as in nature, there is no such criterion of felt necessity. If Newton could not help but believe space and time were absolute, that subjective fact did not make the absolute theory of space and time true, and if a fascist or communist can't help but believe social problems must be solved according to the dictates of his party leaders, that psychological compulsion does not justify their programs. To say, as Holmes did, that "can't help" but believe means "truth to me" does not dispose of the problem of how truth can also be what gets accepted in the competition among people's ideas of *truth to them,* unless it is to be sheer strength in suppressing all criticism and opposition that constitutes truth. The great dissenter did not advocate such antidemocratic suppression of ideas, though much of his evolutionary realism seems to imply yielding to the strongest power. Thus there is no consistent position in Holmes as to what constitutes the strongest power: is it one's own feeling of what one can't help but believe; is it the force of social custom and convention, which Holmes early regarded as more potent than the written law in many cases; or finally, is it the cosmic forces of which we are mere infinitesimal puppets? All three of these compelling powers, the can't helps of the self, the public force of custom and convention, the cosmos in its infinite perspective, received their due and eloquently caparisoned expression in Holmes's writings, but he never resolved their conflicting claims. Nor can these three different grounds of truth, the subjective, the social, and the metaphysical, be reconciled so long as each is made the isolated and exclusive expression of scattered literary attempts to catch the truth in an arresting aphorism painted on a fluttering pennant. Though he had shown his logical strength in youthful discussions with William James about Kantianism and positivism, Holmes had left systematic philosophy behind him after college and the Civil War, in order to embark heroically on the romantic quest of living greatly in the law, "to wreak his heart out after the unattainable." He found the world did not give a straw for ideals, but the quest was an exciting one and its own reward. Besides, he had certainly added his philosophical flutter to the otherwise unromantic world of legal practice:

To the lover of the law . . . no less a history will suffice than that of the moral life of this race . . . the unfolding panorama of man's destiny upon this earth. Nor will his task be done until, by the farthest stretch of human imagination, he has seen with his eyes the birth and growth of society, and by the farthest stretch of reason he has understood the philosophy of its being.[50]

CHAPTER NINE

The Philosophical Legacy of the Founders of Pragmatism

What philosophical fruit can we cull from our historical review of the crossing of arguments about the interpretation of evolution among the members of the Metaphysical Club? Though the Club was only an informal gathering of minds of diverse interests, they were brought together in the university town of Cambridge, Massachusetts during the early 1870's by a common philosophical concern with the tremendous impact of evolution on scientific and social thinking. They applied the Darwinian ideas of chance variations and natural selection to a host of important questions in logic, physics, psychology, history, jurisprudence, and social ethics, and emerged with a new, important pragmatic reconstruction of traditional philosophy. They brought philosophy down to earth and put it to work on the problems of men in the sciences and in the broader cultural changes wrought by the applications of scientific ideas and methods. By making philosophy a useful instrument in the blazing of new paths in the pursuit of truth and justice, they put American thought in the forefront of intellectual and social progress. Their intellectual reactions to evolution were marked by a farsighted, experimental attitude which freed thought from the incubus of theological dogma, authoritarianism, and a priori rationalism. In short, as a result of their studies American liberalism came to philosophic maturity. The common dynamic leitmotiv of their discussions of the meaning of evolution for man was their deep respect for the inviolable, creative character of individual freedom.

In the foregoing chapters the individual differences in accounts and interpretations of evolution show that our American pragmatists did not subscribe to a single coherent all-inclusive system of reality or history. In this concluding chapter I wish to indicate summarily the *common* features of their method of thinking which constitute their legacy to twentieth-century philosophy. For this purpose, I shall deliberately overlook their differences. The fruits of evolutionary pragmatism have borne several component ideas as the seeds of pragmatism today. American pragmatism has fostered, first, an *empirical* respect for the complexity of existence requiring a *plurality* of concepts to do justice to the diverse problems of mankind in its evolutionary struggles. Secondly, it has abandoned the eternal as an absolute frame of reference for thought, and emphasized the ineluctable pervasiveness of *temporal* change in the natures of things. Thirdly, it has regarded the natures of things, as known and appraised by men, to be *relative* to the categories and standards of the minds that have evolved modes of knowing and evaluating objects. Fourthly, it has insisted on the *contingency* and precariousness of the mind's interactions with the physical and social environment, so that even in the most successful results of hard gained experimental knowledge, what we attain is *fallible*. Finally, American pragmatism upholds the *democratic freedom of the individual* inquirer and appraiser as an indispensable condition for progress in the future evolution of science and society.

Though each of these component principles of pragmatism has had a history antedating American philosophy, it is the way they merged and fused together in the thinking of the founders of pragmatism that makes their manner of philosophizing so distinctive. My brief summary and analysis of these seminal principles will be abstracted from the richer texts and contexts previously considered, but with the primary purpose here of suggesting what still lives of the pragmatists' thinking and its continuity with the foundations of the natural and social sciences of our day.

1. PLURALISTIC EMPIRICISM

Pluralistic empiricism is the piecemeal analysis of the diverse issues pertaining to physical, biological, psychological, linguistic, and social problems which resist resolution by a single metaphysical formula. It is the method which Chauncey Wright took a leading part in advocating when he defended Darwin's theory of natural selection by a minute analysis of the observed order and distribution of leaves around the stem of a plant (phyllotaxis), and when he argued that the evolutionary hypothesis did not apply to geology and the cosmic weather of astronomy. Wright's pluralism appeared in the distinctions he made concerning our scientific, ethical, aesthetic, and religious interests against traditional philosophical attempts to fuse them. It is the operational method which Charles Peirce showed was the way to make our ideas clear: consider what conceivable experimental *effects* the objects of these ideas would have. Experimentation becomes a rational method of adapting our ideas to an unknown but not unknowable environment by transforming it in relation to our needs. It is the heart of William James's radical empiricism in which the flux of experience or stream of thought is regarded as marked by singular, irreducible, and genuinely novel, saltatory pulses which cannot be explained away by any single mechanical law such as Spencer's or by a spiritual monism such as Hegel's. It appears even in John Fiske's positivistic quest for evidence of evolution in so many diverse fields: astronomy, geology, embryology, philology, anthropology, and American history, underneath his Christian faith in the spiritual goal of all evolutionary processes. It was the *logica utens* or analytical tool of the lawyer-philosophers, Green and Holmes, when they took pragmatic cognizance of the evolutionary adaptation of the law to social changes.

The increased specialization of the various branches of the sciences today requires the use of this piecemeal method of approach, which is essentially in accord with Aristotle's advice to adapt method to subject matter. Astronomers, geologists, biologists, anthropologists, sociologists, and historians today are wary of beginning with a single a priori scheme of evolution. Only on the basis of years of extensive special researches by many biologists does a scientist like Julian Huxley propose a synthesis under the name of evolution, continuing Darwinian ideas but modifying them seriously in the light of post-Darwinian empirical discoveries. Similar modifications have been made of evolutionary conceptions—always regarded as hypotheses rather than metaphysically binding formulas—in the study of stars, rocks, minds, primitive societies, and modern institutions.

In contemporary philosophy there are various schools that adopt the method of pluralistic empiricism in one form or another: pragmatists, critical realists, logical positivists, and the British Cambridge school of logical analysts. All these schools are disposed to abandon system building and synoptic truth for the piecemeal study of the basic concepts, procedures, and language of the sciences. They have made much progress in clarifying the ideas of truth, causality, probability, meaning, values, and the methods of verification and deduction in sciences as well as in everyday reasoning.

The principle of verifiability as the test of the meaning of an idea was advanced by the founders of pragmatism against the metaphysical assumptions of empirically unverifiable, uncognizable "realities," supernaturally revealed higher or eternal truth, and uncritically held common-sense intuitions of the Scottish school. Our pragmatists held that the materials of human knowledge are provided by experiences originating from the impact of external objects or from the internal operations of our minds as we reflect on the relations involved. They went beyond the "experiential" theory of British sensationalism by advancing an evolutionary conception of objects and of the conceptions used to understand them. Their evolutionary empiricism was thus able to overcome the static character of experience entailed by the passive ideas, sensations, or impressions of Locke, Berkeley, or Hume.

When James was accused of harboring a view of experience that led to subjectivism, he repeatedly insisted that experience had its external aspects—"ex-perience," he would spell it—and that he never intended to deny the realistic impact of "hard data" external to our feelings. Though it does not seem that James consistently worked out a realistic view, there is no doubt that he argued effectively against the monistic doctrine of internal relations which denies the independence and plurality of qualities and individuals. James did join Wright and Peirce in advocating a "critical common-sense realism," but they each meant somewhat different things by that phrase. For Wright, the main task of philosophy was to exorcise critically the vagueness of common sense without abandoning all the practical advantages of common-sense beliefs. Peirce expressed his critical common-sensism as follows:

While it is possible that propositions that really are indubitable, *for the time being*, should nevertheless be false, yet in so far as we do not doubt a proposition we cannot but regard it as perfectly true and perfectly certain; that while holding certain propositions to be each individually perfectly certain, we may and ought to think it likely that some one of them, if not more, is false. This is the doctrine of Critical Common-sensism, and the present pertinency of it is that a pragmaticist, to be consistent, is obliged to embrace it.[1]

The belief that the earth is stationary was sound common sense *while* there was no astronomical evidence to conflict with the deliverance of the senses. Now we may say today either that that proposition was always false or that it was pragmatically believed to be true and is not so any longer. The first of these alternatives commits us to an eternalistic theory of empirical truth, whereas the pragmatic view rejects this theory in favor of a view of empirical truth as *temporal, relative*, and *probable* or *fallibilistic*. Without attempting to justify or defend this pragmatic theory adequately, I shall proceed to outline what each of these characteristics of pragmatic truth meant to the founders of pragmatism.

2. TEMPORALISM

The Heraclitean idea πάντα ῥεῖ runs counter to the ancient notion of the great chain of being in the history of philosophy until we come to "the temporalizing of the chain of being," as Professor Lovejoy has so aptly put it, in eighteenth-century theories of progress and evolutionism.[2] Kant and the post-Kantian *Naturphilosophen* still confined time within the restricted domain of the form of phenomena and left ultimate reality in an eternal realm protected from the vicissitudes of change. This ontological dualism made it possible to conceive of the natures or ideas of species, if not the existent species themselves, as eternal links in the chain of being. By conceiving both the forms of thought and natures of things as themselves products of the flux of evolution, the founders of pragmatism used temporalism to invade the eternal citadel of one of the oldest of metaphysical traditions.

For Chauncey Wright, time was the very substance of physical and psychological phenomena whose complexity produced what he liked to call "cosmic weather." For Charles Peirce, plenitude was to be found in the plethora of logical relations and tychistic possibilities of a universe whose very laws were subject to temporal change; his fundamental doctrine of *synechism* was a logicized version of temporal continuity. For William James, plenitude was to be found in the psychological resources and energies of individuals, and continuity in the stream of consciousness. John Fiske temporalized the mechanism of cultural progress in his theory of prolonged infancy. Holmes shared with Green and Fiske the *historical* approach to the law as an anthropological document.

Critics of pragmatism have argued that knowledge of the past confutes its experimental principle of verifiability, for we cannot verify or transform experimentally what is irretrievably past. The pragmatic reply to this argument refers us to the method of making historical propositions, namely, we use hypotheses and inferences based on *present* evidence (documents, remains, memories of witnesses) and on the continuity of processes *now* observable (erosion, natural selection, population changes), with those of the past. James went very far, indeed, in making truth itself evolve in time. Peirce was more logical in distinguishing truth from our conceptions of it. Evolution applies to our opinions which *would* converge on the truth if an indefinite community investigate long enough.

This pragmatic temporalism leads to a more empirical view of history and knowledge than that which finds eternal laws of development in social change and science. An a priori historicism lies behind the great use made of the comparative or genetic method in nineteenth-century philosophies of history (Hegel, Marx, Spencer). Wright eschewed such evolutionary philosophies of history as forms of a priori "German Darwinism." He also showed the limits of Sir Henry Sumner Maine's historical approach to jurisprudence, an approach which influenced greatly his lawyer friends, Fiske, Green, Warner and Holmes. It is true that Peirce was often addicted to dubious features of historicism in his logical theory of abduction and induction and in his metaphysic of evolutionary love, making much greater use of the genetic method than his more rigorous mathematical logic should have permitted. We have seen how James vigorously opposed Spencerian evolutionism for attempting to submerge great individuals of history in mechanical environmental laws. James favored Darwin's theory of unpredictable chance or spontaneous variations as the unfathomed source of history's great men. He also used the genetic method in describing the case histories of religious experience; but, in the end, we find James sacrificing logic for the sake of even a "crass supernaturalism." Thus James's psychologism may be regarded as a hybrid crossing of individualism and temporalism; in regarding differences of doctrine as arising out of differences of personal temperament in philosophers, he declined to consider the historical influence of previous doctrines as conclusively shaping the course of the history of philosophy. The Hegelian and Marxian dialectic of history never took firm hold of any of the early American pragmatists. In the early 1880's John Dewey was a Hegelian, under the influence of George Sylvester Morris, at the same time that Peirce was teaching scientific logic at the Johns Hopkins University. It was not long, however, before the experimental psychology of G. Stanley Hall, the dynamic idealism of Alfred Lloyd, and the antiformal individualism of William James and Oliver Wendell Holmes, Jr., led Dewey to drop the Hegelian strait jacket of dialectical formalism and to choose the more fruitful method of the natural sciences as the more reliable instrument of inquiry.[3] The geneticism of late nineteenth-century evolutionary pragmatism appeared in Dewey's *The Study of Ethics: a Syllabus* (1894) and "The Evolutionary Method Applied to Morality" (1902). Dewey argued for the view that we cannot adequately explain things or justify moral ideas without knowledge of the psychological and *social processes* or *means* which occur or are made to occur in the production of things or ideals.

The practical lawyers, Green, Warner, and Holmes, did not find the historical and scientific approach to law useful if preached as a gospel of inevitable social progress in Fiske's optimistic fashion. More realistically, they regarded the law as a great anthropological document containing the evidence of man's fumbling for adjustment to social changes amid "survivals" of outmoded customs, jumbled rules, and inadequately defined terms. We must pull the dragon of history out of its dark cave, Holmes would say, in order to see its claws more clearly. The analytical-historical investigations of Green, Warner, and Holmes paved the way for the functional sociological realism of twentieth-century American jurisprudence. This school still aims to adapt the law to the changing conditions of society in order to minimize conflicting group interests and maximize their productive coöperation.

3. RELATIVISM

The nineteenth century saw the breakdown of Europeocentrism; the notion that only the civilization of Europe could produce the best in art, science, and philosophy was challenged by the growing independence and self-reliance of American writers, scientists, and philosophers. Cultural anthropologists (Maine, Tylor, Spencer, and so on) showed Fiske and our lawyer-philosophers—with much more historical data than were available to Locke when he attacked innate ideas—how different standards of right and wrong worked in communities with different cultures.

In psychology and epistemology, experimental discovery of the relativity of perception of qualities, spatial relations, threshold limits, memory span, judgments of discrimination, and so on, accompanied a revival of Berkeleyan and Humean sensationalism. Wright and Stallo, both admired by Fiske in his positivistic relapses, favored the nominalistic view of Mach and Mill that scientific laws are correlations of sensations. Peirce, however, stood out as the chief antagonist of this nominalistic view, and proposed (with Abbot) a modified Scotistic realism of objective logical relations. But both psychological relativism and the Peircean logic of relations aided in the revision of the Aristotelian subject-predicate logic and in the rejection of the absolutistic ontology of Schelling and Hegel. The categories of thought were thrown into the evolutionary cauldron to emerge as flexible relational forms.

The Darwinian theory of natural selection showed that the essential properties of living things depended on the emergence of traits that increased the power of individuals to cope with a hostile environment. Food was more easily obtained when animals hunted for it together; and other joint enterprises, associated with sex at first perhaps, would be furthered by adapting some inherited trait of gesture or voice to develop the rudiments of language. It is characteristic of our evolutionary pragmatists to explain by reference to physiology and language the evolution of the highest reaches of human thought—even in the obscure and foggy region of "self-consciousness." Whence another important sense of relativism: namely, the relativity of mind to the physiological, social, and linguistic conditions of human behavior. This view of the mind borders on the method of behaviorism: the mind is to be understood and mental phenomena predicted according to the observable phenomena of behavior, physiological and linguistic. The pragmatic meaning of any expression will then be relative to the context of such behavior. If a man raises his hand to me or shouts at me, his intention or meaning can be inferred from antecedent actions or words accompanying similar gestures or words. His flushed face, dilated nostrils, tense inflections of voice, and so on, are behavioral expressions of anger, unless we know that he is an actor who has mastered the art of imitating these external signs of anger. Darwin's study of *Expressions of Emotion in Animals* and William James's physiological theory of the emotions prepared the way for twentieth-century behaviorist psychologies. Generalize the behavioristic method of approach and you have a contextualistic or relativistic theory of meaning: *the meaning of a statement varies with the spatiotemporal, linguistic, or socio-psychological conditions of its occurrence.* This generalization is in line with subsequently developed ideas of relativity in physics, logic, psychology, and sociology.

When Holmes defined truth as what he could not help but believe, he was a *psychological* relativist. So also must we classify James in his will-to-believe doctrine. Wright and Peirce were *logical* relativists, Fiske was a *historical* relativist, Green, Holmes, and Warner were *sociological* relativists: the law is what the courts decide in different social contexts. F. E. Abbot and Peirce thought of truth as directed towards an indefinite community of minds and an evolving realm of *objective* relations; hence, we may consider their relativism to be *metaphysical*. There will be as many *relativisms* as there are different theories as to what the objects dealt with are relative to.

Relativism in pragmatic ethics, which makes an action good if its consequences are desirable under specific but variable conditions, does *not* imply that there are no valid moral distinctions or principles beyond one's arbitrarily expressed desires. An action is pragmatically justified not because it is subjectively approved, but because the beneficial consequences for all concerned outweigh the harmful ones within a given concrete situation. What the consequences of our actions are is an objective question; whether these consequences conform to what is beneficial to the greatest number of people, and what *is* thus beneficial are also objective questions, though it is often difficult to ascertain in both cases just what the consequences or benefits are. Traditional theological and metaphysical ethics claim universality and certainty for their inviolable rules, eternal commandments, and categorical imperatives. There is a parallel between relativistic theories of knowledge and pragmatic or utilitarian ethics. In both cases, eternal and infallible rules are replaced by contextual, empirically tested generalizations as probable guides, subject to revision in the light of observable agreement or disagreement with predicted consequences. It is relative to the frame of reference of an observer sitting in a train that his judgment that another train is moving makes sense and can be tested; it is relative to what is desirable for individuals in a given society at a given time that what happens in that society is good or bad. Ethical relativism regards value judgments as empirical hypotheses, some more fully confirmed by experience than others; for example, telling the truth or keeping promises leads to one's usefulness and acceptance as a reliable person in most civilized societies. The dependence of all ethical rules on their instrumentality in serving the needs and interests of the greatest number of people makes it necessary to adapt these rules to the changing interests and individual differences of people. Though man is not the measure of all things in the evolutionary process, for they are not all within his control, he is the measure of what is judged good, for that judging is entirely relative to our ideals of what is humanly desirable. Our New England pragmatists were brought up on an Emersonian faith in man's ability to live nobly in accordance with an eternal ideal that was supposed to transcend the competition of the market and the struggle for survival. But they soon realized that Emerson's ideal was too remote from the competitive world of action. They realized that ideals had to be fought for in the midst of an evolutionary struggle, with due regard for both a hostile environment and for conflicts among men. Experience and time reveal the dependence of realizable theoretical and ethical constructions on their relations to external circumstances. We are never certain that we know all these circumstances, but we can be practically confident that we know some of them by means of hypotheses accepted as probable to the extent that they have been empirically tested. Subjectivism and fruitless skepticism are more likely to occur with the absolutist who makes his private intuitions the tests of truth and value, rather than with the pragmatic relativist who regards all such claims as hypotheses in need of public confirmation.[4]

4. PROBABILISM AND FALLIBILISM

The terms "Probabilism" and "Fallibilism" refer to the abandonment of mechanical determinism in physical and social sciences by viewing their laws as probable or contingent. There were two different sources for this twentieth-century idea among our earlier pragmatists:

(a) The metaphysical notion handed down through the ages from Plato to Leibnitz and Hegel that all empirical knowledge is unsatisfactory and inadequate because of the contingency and limitations of sense experience on which "merely" empirical science is based. Certainty is attainable only in the a priori truths of logic, mathematics, and metaphysics. Peirce, Fiske, and James often accepted this metaphysical ground for abandoning certainty in physical, historical, and psychological sciences, but they also argued strongly against a priori absolutistic metaphysics.

(b) The pluralistic empiricism or contextualism described in the above pages would make it logically *unnecessary* for a modern pragmatist to assume that whatever was true of large-scale or macroscopic phenomena, for example, would hold for each individual element in the small-scale or microscopic field of observation. Of all the founders of pragmatism, Peirce had the clearest apprehension of this statistical principle, largely as a result of his scientific work with the United States Coast and Geodetic Survey, though he yielded also to (a) in his metaphysical moods. James accepted Peirce's tychism as a support for his own tender-minded metaphysical individualism and indeterminism. Wright, Green, and Holmes regarded all our empirical knowledge as contingent and fallible in both the natural and social sciences.

Against the metaphysical claims of (a) they argued that in these sciences there was no knowledge that transcended the uncertainties of empirical evidence. They found with Locke that the candle that glows dimly within us is bright enough for all practical purposes. Practical problems require us to know *what is true for the most part*. The latter phrase, taken from Locke, expresses roughly what we mean by probability as "the guide of life."

Though Hume had dealt a mortal blow to alleged proofs of necessity in factual matters and insisted on probability as our only empirical guide, he still hankered for practical or "moral" certainty. He thought he found it in the psychological compulsion of habit or custom as the universal law of association, comparable to Newton's law of gravitation. Wright in his skeptical theories of scientific neutrality and metaphysical nihilism, Green in his Benthamism, and Holmes in his "bet-abilitarianism," seem to have adhered more closely to Hume's psychological probabilism than James, Fiske, or Peirce. None but Peirce explored the logical meaning of probability, and followed J. Venn in the formulation of an objective frequency theory of probability, however incomplete and unsatisfactory it is from the standpoint of more recent rigorous logical studies. James and Peirce seemed to regard the fallible character of scientific laws as providing loopholes for the agency of spiritual forces in human nature. Apart from these divergent directions of their metaphysical speculations, the founders of pragmatism welcomed Darwin's evidence of the role of individual variations in the evolution of living species. Peirce called attention to the impending breakdown of classical mechanical laws in statistical physics, and tried to generalize evolution as a theorem in the logic of probabilities. What remains of these speculations concerning probability is the abandonment of certainty in both the natural and social sciences and the acceptance of an empirical method of testing generalizations about either natural or social phenomena. The neo-Kantian separation of *Naturwissenschaften* from *Geisteswissenschaften* was supposed to have bestowed on psychical (social and cultural) phenomena a "higher" certainty than "merely" natural empirical science could possess. The materialistic *Naturphilosophen* (Oken, Büchner, Feuerbach, Haeckel) did not deny the necessity of sharply separating physical from psychical and social phenomena, but simply inverted the order of dependence. What distinguishes pragmatic naturalism from both spiritualistic and materialistic philosophies is the denial of any necessary or privileged status for the findings or "laws" of either psychical or physical phenomena. Neither are demonstrably certain if the nature of any generalization is that of a prediction or hypothesis confirmable to some degree. Only experience can furnish the neutral materials for physical or cultural generalizations. These materials are temporal and complex, hence not deducible from any favored eternal and single system of timeless or simple entities dwelling in realms of matter or spirit. The plurality of events and qualities experienced in natural or human history requires recognition of the tentative, probable, and fallible character of all empirical knowledge. The unique individual as such eludes all formulations whether scientific or historical. In experience we deal always with individuals, and in scientific formulations with their classifiable characters. Now pragmatism looks to the scientific method for selecting those classifications. In so far as that method is simply critical common sense, it should be always ready to correct itself, knowing that its findings are only probable, or confirmed only to a degree.

5. SECULAR DEMOCRATIC INDIVIDUALISM

The fallibility of the most exact scientific findings, so often stressed by the founders of pragmatism, was part of their profound liberal aversion to all forms of dogmatism, including that of overzealous lovers of specialized scientific truth. They all objected to the hierarchical priesthood of social scientists advocated by Auguste Comte in his later years, for the same democratic reason that led them to reject theological and metaphysical dogmas. They opposed any privileged class of revelations, and all authoritarian educational and political systems. The founders of pragmatism had diverse interpretations of the relation of religion to science. Peirce, like Abbot and Fiske, was an evolutionary theist; James rebelled against evolutionary theology and agnosticism by advocating a "crass supernaturalism" on the basis of the variety of religious experiences he had empirically investigated; Wright, Green, and Holmes regarded science as entirely indifferent or neutral to religious issues. Like Emerson they looked upon religion as each person's own affair; let each decide to what final tribunal he wishes to submit his way of living. Hence, despite their diverse interpretations of religion, they agreed in recognizing its historical, social, and ethical function, and they opposed religious intolerance, ecclesiastical authoritarianism, and intervention in political or educational institutions. They opposed slavery in the southern plantations and the exploitation of labor in the northern mills. Godkin's *Nation* was their favorite liberal organ. William James fervidly opposed the growing imperialism of our government at the time of the Spanish-American war. These are only brief indications of the democratic manifestations of their pragmatic liberalism.

This heroic individualism of the founders of pragmatism is seen in Nicholas St. John Green's defiance of *stare decisis,* in Chauncey Wright's disdain for theological systems of morals or of science, in Charles Peirce's sweeping condemnation of all modern philosophies as tainted with nominalism, in William James's challenge to agnosticism in himself as well as in others with his will to believe, and in Oliver Wendell Holmes's twisting of the tail of the cosmos. They were each too independent and strong as thinkers to club together as a school in defense of a system. They were too keenly aware of the tyrannical way in which such schools or systems of thought dominate and obstruct freedom of inquiry. So each was content to forge his way through the stubborn prejudices and too rigidly fixed authorities in his own field of study.

The common political faith shared by all our early pragmatic thinkers was based on a utilitarian and democratic ethic of individualism to which all social institutions were subservient. It is a travesty on American pragmatism to condemn its philosophy as crass opportunism, as subordinating truth to cash value, or as the ideology of "Wall Street imperialists." The principle of utilitarianism, "the greatest happiness of the greatest number," is the ethical basis of democracy, and our American pragmatists clung to this principle with a truly religious fervor, against the rugged individualism of "social Darwinists" like Spencer, and against the ruthless collectivism of the desperate European revolutionists of 1848 and 1870. It was possible for pluralistic pragmatism to treat man's social sentiments as capable of evolving beyond the animal basis of tribal gregariousness and of transcending, by the slow process of education and social evolution, the brute level of the biological struggle for existence. Thomas Huxley's essay on "Ethics and Evolution" pitted man's enlightened moral will against the indifferent cruelty of cosmic evolution in the same humane way as did our American evolutionary pragmatists in their ethical concern. James was even inclined to belittle science in order to uphold this faith, and appealed to a secular and democratic philosophy of religion against a trivial-minded aimless life with no motive but self-interest or the pleasure of the moment, achieved at the expense of other's labors. The high ethical quality of their diverse conceptions of evolution stood opposed to the harsh competitive character and ruthless self-aggrandizement which "social Darwinism" tried to justify in the growing commercial expansion of the United States. On the whole, there was much more interest in social ethics than in applied economic science among our thinkers, but their lack of such knowledge does not diminish the evidence of their aversion to materialistic opportunism of either an egotistic or totalitarian variety.

The challenging question of the post-war world is what sort of world-wide institutions and practical guidance civilization may expect from American thinkers trained in scientific methods as means for realizing democratic ideals. I have indicated that paths were broken in a historical instance of this problem by a group of American philosophers weathering the intellectual storm of the Darwinian controversy by freeing the natural and social sciences from the dead hand of closed metaphysical and theological systems. The habits of disinterested pursuit of truth, with respect for, but unyielding critical analysis of, past and prevalent opinions, was coupled with a deep religious respect for individuals and a sense of the practical need for scientific and philosophic enlightenment. No future civilization can dispense with that kind of secular religion and political respect for civil liberties.

If American philosophy is to continue to be a significant cultural force in the world, it will have to draw on its pragmatic legacy. That legacy contains the reasoned and humanitarian faith that the future course of evolution has room in it for the coöperative efforts of free individuals to enrich life with peaceful and creative activities beyond the sheer struggle for existence and power.

APPENDIX

A

Biographical Notes on Chauncey Wright (1830–1875)

Chauncey Wright was one of nine children of Elizabeth Boleyn (or Bullen) and Ansel Wright, grocer and deputy sheriff of Northampton, Massachusetts. Chauncey's grandfather had been a revolutionary soldier, and might have heard the fiery sermons of that other celebrated Northampton philosopher, Jonathan Edwards. The first American ancestor was Samuel Wright, who had come to Boston from England in 1630 and was one of the first settlers of Northampton in 1654.

Northampton was also the birthplace of Chauncey's Harvard friends, his biographer James Bradley Thayer, and Dean E. W. Gurney. There also he met Nicholas St. John Green, with whom he was prepared by the Rev. Rufus Ellis (of the little Unitarian church which Chauncey's father constantly attended) to meet the Latin and Greek requirements for admission to Harvard. Chauncey wrote something of his "Life" for the alumni class book in 1858, six years after graduation, in which he reports that his father was a Democrat and an ardent supporter of Andrew Jackson. An inkling of Chauncey's childhood and temperament appears in this autobiographical sketch:

From the earliest period of my conscious life, I have shrunk from everything of a startling or dramatic character. I was indisposed to active exercise, to any kind of excitement or change. I was never remarkable at any kind of sport; never could see the value or significance of any kind of formality. In illustration of this, I remember a circumstance in my earliest school-days. The school-mistress wished to introduce the custom of kneeling at morning prayer. This I obstinately refused to do, or at least obstinately did not do. The penalty for my disobedience was to kneel by the teacher's side,—a position of dramatic interest by which my spirit was broken.

I was in general a very tractable boy, and never was flogged at school, though I remember some corrections. I had some little ambitions, such as all boys have, but they were for the most part of a solitary nature. I never aspired to be a leader among boys, and never cared for their quarrels and parties. If I aspired to a place, it was to a solitary place and a peculiar one, not within the general aim of the boys. At one time, however my ambition took a social turn. While I was still in the district school, I conceived an ardent attachment for one of the school-girls, which I have never mentioned before this writing to any living soul. I did not even intimate to the young lady herself, but rather built small castles, or very diminutive houses in the air, wherein I dwelt in fancy with her I fancied,—I will not say adored. Such was the character of all the attachments or fancies I have subsequently had . . .

I had in my boyhood a violent temper; but I was not quarrelsome, nor did I ever cultivate pugnacious qualities. My indolence has since completely mastered my temper.

At the age of ten, I formed a liking for the study of astronomy, and my zeal in this easily overcame all the fondness for those surprises which had previously constituted the attraction of New Year's gifts, and so asked my father for that puerile book Burritt's Geography and Atlas of the Heavens; but I afterwards lost my interest in this study.[1]

The future biographer of Chauncey Wright will perhaps tell us what kind of a school it was at Northampton that could make him kneel in prayer against his will and also employ a teacher of natural history, a David Sheldon, who could inspire him to take a profound interest in natural sciences. One would like to know more about the religious habits of the New England community of his youthful years which might throw some light on Chauncey's later antipathy to theological creeds. The standard biographies do not tell us how he became transformed from a boy "much given to solitude and inclined to melancholy" to a man who was "neither solitary nor unsocial"?[2] The personal letters and high testimonials of many friends indicate a very amiable and genial person. He often entertained children with sleight-of-hand tricks, and soothed sick friends.

At Harvard College, which he entered in 1848, after overcoming his defective knowledge of Greek, we are told that he remained "poor in languages but brilliant in mathematics, natural sciences, and philosophy." Yet the theory of the origin and function of language played an important role in his correspondence with Darwin and in his major article on "The Evolution of Self-Consciousness." We know something about the scientific books he read—Dana's geology, Bacon, Emerson, Hamilton, and Mill in Philosophy —and the teachers he had at Harvard—Benjamin Peirce in mathematics, Asa Gray in botany, Jeffries Wyman in physiology. After his graduation in 1852 and until 1870 he worked as a computer for the *Nautical Almanac* office in Cambridge, and wrote many reviews for the *North American Review* and the *Nation*.

What led him to become so much interested in the "mental sciences" that he gave up his work as a mathematical astronomer to lecture on psychology and to review philosophical books instead? Why was he never invited to a permanent teaching position at Harvard, where he was so highly re-

---

[1] *Letters of Chauncey Wright, with some account of his Life*, edited by James Bradley Thayer (Cambridge, Mass., 1878), pp. 5ff.

[2] Cf. J. B. Thayer (*ibid.*) for the latter characterization, and E. S. Bates for the former (*DAB*, article "Wright, Chauncey").

garded by James, Peirce, Charles Eliot Norton, Dean Gurney, Asa Gray, and others? Was it because of his antitheological views, or, as in the case of Peirce, because of his "irregular habits"? [3] Or because, as Fiske and Warner testify, he was such a poor public lecturer? President Eliot's letter to Wright expressed regret that Wright's university lectureship in mechanics (1874) was not to be renewed because of poor registration. Yet Wright, unsuccessful as a classroom teacher, showed a great interest in the theory of education. He criticized the formal discipline theory of the value of the study of classical languages and mathematics.

Wright's views on education were modern and progressive. He opposed the formal discipline theory as an unfactual justification of the value of mathematics and the classics through the unproved transfer of habits of thinking.[4] This denial of formal transference of discipline from one rigorous subject to other unrelated activities was part of his pluralistic empiricism. In a fragment of an early essay, "The Philosophy of Mother Goose," we find the modern idea that children learn more and work better if their *play* rather than work is supervised. At the time President Eliot of Harvard was introducing the elective system, Wright advocated that the college curriculum should aim at permitting a student to master one or two subjects and choose enough of a smattering of other subjects to realize how little he knew of them. Finally, in his Socratic manner of prolonged and thorough discussion in which the vagueness of ideas could be exorcised, Wright kept philosophy alive not as a formalized subject but as a way of thinking about any subject, as a habit of persistent questioning of fundamental assumptions, as an attitude to the diversity of life's problems that cannot be settled by any single formula or system. The practical necessity of adapting our ideas and institutions to the cosmic weather of change is the basis of his methodology. When we appreciate the educative influence of Wright's philosophical conversations on Peirce, James, Green, Warner, Holmes, and Fiske, and note especially that Wright defended Darwinism as a *method* [5] rather than as a cosmology, it becomes clear that he has earned the historical right to be considered the precursor of the empiricistic and pluralistic varieties of pragmatism.

[3] *Letters of Chauncey Wright*, p. 137: "Irregular habits of work and sleep, physical inactivity, and the practice of excessive smoking, had brought on sleeplessness and physical suffering, from which he sought relief in stimulants."

[4] Chauncey Wright, review of I. Todhunter's *The Conflict of Studies, and other essays in subjects connected with Education*, in the *North American Review* (July 1875), reprinted in *Philosophical Discussions* (New York, 1877), pp. 267–295.

[5] Wright noted that "adaptive" characteristics of organisms ought not to be contrasted "absolutely" with "genetic" ones, and added that the relations of utility prominent in Darwin's discoveries illustrate "the *method* by which the principle of Natural Selection is to be applied as a working hypothesis" (*Philosophical Discussions*, p. 324).

The son of William James who edited his *Letters* not only under-estimates but mis-states the facts concerning Wright's influence on his father:

> It has been suggested that Wright influenced James's thinking. If so, his influence was not lasting, and, in the opinion of the editor, can easily be overstated. James was not limited to any one philosophic companionship even at this time (1869–71); and if he felt Wright's influence, it is remark-able that there should be no mention of him in any of the letters or memo-randa that have survived and that there was never any acknowledgement in James's subsequent writings.[6]

The error of the editor's judgment concerning his father's intellectual debt to Wright is surpassed by the factual mistake concerning the lack of mention of Wright in the letters or subsequent writings of William James, all the more surprising because the second volume of *The Letters of William James* contains the letter (quoted above) which mentions Chauncey Wright, Charles Peirce, St. John Green, Warner, and Fiske, who met with James to discuss Fiske's *Cosmic Philosophy*. Then, in another letter of William James, when he was away on a scientific expedition in Brazil with Louis Agassiz, and "suffering from intellectual as well as domestic nostalgia," we find James saying: "Would that I might hear Chauncey Wright philosophize for one evening." [7] Besides the obituary notice of Chauncey Wright which James wrote for the *Nation* in 1875, there are acknowledgments of Wright's influence in the preface to *The Principles of Psychology* (1890), which records James's indebtedness to Wright and to Charles Peirce "for their intellectual companionship in old times," as well as in his notes on "Religious Guarantee" and "Against Nihilism" published by Professor R. B. Perry.[8] The more serious mistake of the editor of James's *Letters* consists in his characterization of Wright's philosophy as "vigorously materialistic." The groundlessness of this characterization is evident in Wright's defense of the neutrality of science with respect to either spiritualistic or materialis-tic metaphysics.

The distinguished brother of William James has provided us with such a sympathetic and touching picture of Chauncey Wright's place in the intimacy of James's household and intellectual surroundings, that it must be a sheer blind spot that would explain the son's editorial version of Chauncey Wright's relation to his father. Though Professor Perry has quoted the first

---

[6] *The Letters of William James,* edited by his son Henry James, 2 vols. (Boston, 1920), I, 152n.

[7] Ralph Barton Perry, *The Thought and Character of William James,* 2 vols. (Boston: Little, Brown, 1935), I, 520, 221; letter of William to Henry James, July 23, 1865.

[8] Perry, I, 523ff.

part of Henry James's delicate description of Chauncey Wright, I shall cite the whole passage, for it might well have served the great novelist and analyst of the inseparable emotional-intellectual life of the mind as a theme for a biographical novel:

Chauncey Wright sits for me in his customary corner of the deep library sofa and his strange conflicting conscious light blue eyes, appealing across the years from under the splendid arch of his fair head, one of the handsomest for representation of amplitude of thought that it was possible to see, seem to say to me with a softness more aimed at the heart than any alarm or any challenge: "But what then are you going to do for me?" I find myself simply ache, I fear, as almost the only answer to this—beyond his figuring for me as the most wasted and doomed, the biggest at once and the gentlest, of the great intending and unproducing (in anything like the just degree) bachelors of philosophy, bachelors of attitude and of life. And as he so sits, loved and befriended and welcomed, valued and invoked and vainly guarded and infinitely pitied, till the end couldn't but come, he renews that appeal to the old kindness left over, as I may say, and which must be more or less known to all of us, for the good society that was helplessly to miss a right chronicler, and the names of which, so full at the time of their fine sense, were yet to be writ in water. Chauncey Wright, of the great imperfectly-attested mind; Jane Norton, of the train, so markedly, of the distinguished, the sacrificial, devoted; exquisite Mrs. Gurney, of the infallible taste, the beautiful hands and the tragic fate; Gurney himself, for so long Dean of the Faculty at Harvard and trusted judge of all judgments (this latter pair the subject of my father's glance at the tenantship of Shady Hill in the Nortons' absence:) they would delightfully adorn a page and appease a piety that is still athirst if I hadn't let them pass.[9]

When Chauncey Wright died suddenly while working over his writing into the early hours of morning, as was his habit, only a few of his friends were in Cambridge: St. John Green, Hooper, and Henry James. James had gone immediately to Wright's house on hearing of a physician being called, but it was too late.

Mr. James said that about a week ago [another friend of Wright's reports] he spent an evening at his house, and had never known him more delightful, —talking in his best vein, not about things and ideas, but about people, giving astonishingly minute characterizations of them. Innocent, mild, kindly, sympathetic, pure-minded,—these were the words Mr. James used about him; and, in speaking of his relations at the Nortons and his admiration of Miss Jane, he quoted his saying that he would rather have her personal approbation than all the fame of this country and Europe. It was a great satisfaction to hear Mr. James's appreciative talk, and to have all his lovely and amiable ways brought up; and I could not but hope that Mr. James was mistaken, of late at least, in thinking that he had been so miserably unhappy.

    [9] Henry James, *Notes of a Son and Brother* (New York: Charles Scribner's Sons, 1914), pp. 282f.

I had hoped that in the last year or two things had been less difficult with him. He has certainly seemed, as I have seen him, quiet and serene . . . Mr. James said that his face was perfectly white and very noble to look at.[10]

[10] *Letters of Chauncey Wright,* pp. 358f. The following letter of Leslie Stephen to Charles Eliot Norton, May 5, 1877, indicates the British critic's reactions to Wright: "Since I wrote to you last, I have read Mr. Chauncey Wright's book Philosophical Discussions, ed. by Norton, 1877, or nearly all, and—to say the truth—found it a tolerably tough morsel . . . Perhaps I am a little spoilt by article-writing and inclined to value smartness of style too highly. The only point which struck me unpleasantly in the substance of the book was his rather over-contemptuous tone about Spencer and Lewes. I don't doubt that his criticisms of Spencer are tolerably correct, though I can't see that Spencer really means to concede so much to the enemy as C. W. supposes; but I confess that Lewes seems to me to be a remarkably acute metaphysician, and one who will make his mark . . . Anyhow, Wright must be a great loss. Nobody can mistake the soundness and toughness of his intellect, and his thorough honesty of purpose. I had the pleasure the other day of showing the book to the great Darwin, who had already received a copy from you"; F. W. Maitland, *The Life and Letters of Leslie Stephen* (New York, London, 1906), p. 300.

B

The Chauncey Wright Papers at Northampton, Massachusetts

(In possession of Mrs. Dorothy Pearson Abbott, Alumnæ House, Smith College)

Besides the two essays reproduced here, there are eight letters from Charles Darwin whose contents I have described in Chapter III, sect. 3, based on my article in the *Journal of the History of Ideas*, VI, 19–46 (1945). The two essays were probably college themes written about 1850. The future biographer of Chauncey Wright may wish to see also the contract he had from the U. S. Naval Office of the Nautical Almanac to work as a computer after 1852, and the letters from President Charles Eliot appointing him as university lecturer in psychology (1870) and in mechanics (1874), including a petition for the class in mechanics, dated October 27, 1874 and signed by George S. Pine, B. O. Peirce, Edward B. Lefavour, Henry Sargent, Frederick L. Green, H. R. Mills, Richard S. Culbrith, S. T. Fisher, C. T. Peckham, F. C. McDuffee, H. Amory, and A. H. Witherbee.

There is a batch of correspondence marked "Personal" from Grace Ashburner, John Eden, George H. Fisher, Mrs. Forbes, George Frisbee, Erastus Hopkins, C. P. Huntington, James Lyman, Mr. and Mrs. Peter Lesley, Sara Sedgwick, Underwood, and Ansel Wright, Sr.; a letter from Lydia C. Dodge of Hingham, Massachusetts, October 31, 1864, inviting Wright to lecture there on natural science.

The remaining correspondence, arranged alphabetically, with my digest of each letter, is as follows:

From L. C. Agassiz: (1) n.d. Cambridge, The College: change of recitation hour. (2) November 11, —, same: class in natural philosophy formed. (3) June 26, —, same: asks for number of recitations given by Wright (at Agassiz' School for Girls) in order to ascertain salary.

From John Bigelow: (1) November 12, —, Evening Post, New York: introduces John C. River of the Washington *Globe*. (2) August 6, 1860, same: pays C. W. $35 "for several contributions."

From William Boise: June 5, 1850, Yale College: thanks C. W. for two solutions to mathematical problems in Playfair's Euclid.

From Francis Bowen: (1) June 25, 1860, Cambridge: criticizes C. W.'s paper on the economy of design in bee's cells for failing to note that it is "God's work, not the bees!" (2) October 9, 1864, Harvard College: replies to C. W.'s review of Bowen's *Logic* in *North American Review* by defining probability and causality in theological terms *vs.* Darwin. (3) October 19, 1864, same: physics gives only probable knowledge, metaphysics the truly causal knowledge of necessity. (4) October 25, 1864, same: thinks C. W. wrong in regarding the arithmetic mean as valid as the square root mean and even better for certain purposes.

From George L. Cary: (1) November 26, 1854, Medway, Massachusetts: invitation to C. W. to visit. (2) September 15, 1861, Germantown, Pennsylvania: Lincoln "a gorilla in the executive chair."

From Reginald H. Chase: (1) September 13, 1862: no drafting of soldiers in the 22d Ward (Chestnut Hill). (2) August 17, —: Mr. Apple of Philadelphia quoted: "this nigger war." (3) October 30, 1864: on recruiting difficulties.

From Dr. David Cheever of Boston: February 1, 1853, Portsmouth, New Hampshire: "Deutero-scopy."

From George F. Comport: October 26, 1868, 75 E. 10 St., New York City: establishment of American Philological Society.

G. W. Curtis: October 25, 1868, Staten Island, New York: "Dána must be defeated."

Charles W. Eliot: (1) March 15, 1870, Harvard College: appointment as university lecturer on psychology twice a week October 1 to February 11. (2) July 6, 1875, same: Physics 1 discontinued since only one person has elected the course.

W. P. Garrison (editor of the *Nation*): (1) November 4, 1868, 3 Park Place, New York: asks C. W. to review Porter's *On the Human Intellect* and Bain's *Mental Science;* discusses teaching of geometry and algebra. (2) October 28, 1874, 5 Beekman St., New York: asks C. W. to review McCosh's *Scottish Philosophy* and Archer Butler's *Lectures on the History of Ancient Philosophy.*

E. L. Godkin: May 31, 1874, *Nation*, New York: opposes Mill's argument for equality of sexes. Women's submissiveness is ineradicable and hereditary (cf. C. W.'s lengthy reply in *Letters*, pp. 257–265, June 3, 1874).

B. H. Gould: February 25, 1866 and March 14, 1866: complains about printing of his communication to Academy of Arts and Sciences; C. W. replies March 2, 1866, explaining his moral policies and duties as corresponding secretary of the Academy.

Asa Gray: February 13, —, Cambridge, Massachusetts: returns "Fendler's queer book" loaned to him by C. W.

Charles E. Grinnell: December 13, 1873, Charlestown, Canada: reports a subcommittee on philosophy at Harvard, Park, Emerson, Means, and Grinnell to visit in psychology, with Park and Cabot in modern German philosophy, logic, and metaphysics; Emerson and Grinnell also to visit modern German philosophy.

E. W. Gurney: two letters August 1 and 6, 1857 inviting C. W. to visit at Rye Beach.

Edward Everett Hale, editor *Christian Examiner*, Boston: (1) September 24, 1857: asks C. W. to review Blodgett's *Climatology* at $1 per page. (2) November 13, 1857, same: sends payment of $6.75 for reviews.

Catherine E. Ireland: seven letters from Boston and New York beginning July 6, 1864, from a "former student and friend."

William Woolsey Johnson: (1) May 3, —, Annapolis, Maryland: asks C. W. to explain meaning of "mean distance" in astronomy. (2) December 31, 1873, same: asks if there is any job for a mathematician around Cambridge.

W. E. Logan: December 20, 1856, Montreal, Canada: determination of longitude of Montreal.

J. E. Oliver: October 12, 1855, Lynn, Massachusetts: Elijah Pope's phonographic school, Portland. May 1, 1859, same: a mathematical theory of individuality based on C. W.'s theory of phyllotaxis.

John C. Palfrey: February 28, 1874, Lowell, Massachusetts: mathematics committee to visit Harvard.

Charles S. Peirce: September 2, 1865, St. Albans, Vermont: Mill on Hamilton; Card game (cf. Chapter II, note 16, below).

John D. Runkle: (1) August 9, 1862, Dedham, Massachusetts: invitation to visit Pigeon Cove. (2) September 21, 1873, Brookline, Massachusetts: account of payments made to C. W. (Runkle was in charge of the Nautical Almanac Office in Cambridge.)

David S. Sheldon (C. W.'s schoolteacher of science at Northampton): August 25, 1858, Iowa College, Davenport, Iowa: introduces a student, Clemens Hirschl.

Joseph H. Sprague: five letters, 1854–1860 (Northampton, Massachusetts): invitations; borrows $10; asks for job with Nautical Almanac.

James B. Thayer: (1) August 24, 1850, Cambridge, Massachusetts: returns money borrowed from C. W. (2) February 13, 1850, Milton, Massachusetts: talks of Fanny, Hannah, Mrs. Forbes, and Kemble of Northampton. (3) April 8, 1871, Boston: Emerson to be at Fisher's; will C. W. serve as bond on loan of $1000?

William S. Thayer: (1) February 13, 1851, Boston: crack at Francis Bowen. (2) July 28, 1857, New York *Evening Post*: will C. W. write of tornadoes this summer, including their "philosophy"? (3) May 5, 1857, same: sends $10 for C. W.'s recent article in *Post*. (4) August 20, —: will C. W. speedily write article on Atlantic telegraph? (5) September 11, 1853, New York: "Some rowdyish times here at Woman's Rights and Temperance Convention"; Miss Pomeroy recalls two pleasant visits from C. W.; "P. S. There is a woman on exhibition here who weighs 754 lbs. How would you like to embrace a wife like that?"

J. M. Van Vleck: December 10, 1855, Middletown, Connecticut: mathematical studies of Peirce and of Cauchy recalled of undergraduate days at Harvard.

D. E. Ware: (1) November 29, 1852, Baltimore: praises lectures of Orvil Dewey, D.D. (2) August 6, 1862, Boston: invites C. W. and Gurney to take excursion.

W. R. Ware: (1) June 6, —, Milton, Massachusetts: invitation to C. W. and Gurney to visit. (2) June 15, 1864, Boston: in army hospital.

D. P. Wilder: (1) May 13, 1852, Baltimore: asks C. W. to solve a problem in trigonometry from Davies' Legendre. (2) September 27, 1852, same: teaching algebra and geometry and expecting to teach trigonometry; "Do you intend to study any profession? Or do you like math so well that you shall dig into them and finally take 'Benny's' [Benjamin Peirce's] place?" (3) February 27, 1853: asks C. W. to solve problem in algebra from Bourden.

There is also among the papers a "Positivist Hymn on Water."

Of the hundred letters to and several papers by Chauncey Wright at Northampton, only parts of eight letters from Charles Darwin have been published. Indicative of Wright's early interest in ethical as well as scientific questions are two undated essays, each several pages long, marked "For Ansel" (Chauncey's brother), which read like college freshman themes: "Whether the government of this country ought to interfere in the politics of Europe to aid or countenance those who are struggling for liberty there?" and "Was the act of Brutus justified in killing Caesar?" Wright's affirmative answer to both questions seems to be in youthful but dignified imitation of Mill's abstract style and humanitarian utilitarianism. We must not lose sight of the ethical interests of Wright even though he constantly distinguished ethical principles from those of natural sciences and metaphysical theology.

Two Essays by Chauncey Wright

1. WHETHER THE GOVERNMENT OF THIS COUNTRY OUGHT TO INTERFERE IN THE POLITICS OF EUROPE, TO AID OR COUNTENANCE THOSE WHO ARE STRUGGLING THERE FOR LIBERTY?

When charity and gratitude are enjoined as general duties, it is not meant that an unlimited and indiscriminate exercise of kindness and gratitude may not interfere with other duties and thus become not only useless but pernicious; so in discussing the duty of intervention we are not going to maintain that there are not kinds of intervention that are wrong or that there are not many kinds of intervention which in particular crises may be positively injurious; but simply that intervention in general is the duty of the governments of particular nations in behalf of universal humanity, and that there is nothing which frees our government from this universal obligation of oppressed humanity.

We are not to conclude from the notion of governments, as many are disposed to do, that the whole range of its duties extend merely to fulfilling the obligations it owes to the people expressed in its constitution or by its general usages, any more than the whole duty of an individual consists in doing what he has bound himself to do for others, and in not interfering with the rights of others; for the moral law is so universal in its application that it requires every freely acting agent, whether it be single or compound to do all that lies in its powers or privileges towards the accomplishments of moral ends.

There are many acts neither belonging to the common duties of governments nor yet interfering with these duties, but lying within the power of government to do; which, if they are right, the government is bound to do.

If then intervention in general behalf of the rights of humanity does not necessarily interfere with the obligations of our government to our people or to other governments, it is required by morality. Protests in general are not injurious to the peace of the world, do not interfere with the obligations of governments to their people or to governments, with which they have formed treaties; are in many cases useful, and therefore in these cases duties. They do not require further intervention, and so do not lead to fatal consequences in that respect.

Some may however deny that protests can be of any value unless followed up by armed intervention or unless they imply that an armed intervention is to follow if they do not succeed, since the only way in which they effect anything is through fear. But if there is anything (more than triteness) in the saying that opinions are often more efficacious than armies; then to awaken fear is not the only object of a protest. In fact, there are great and important uses in protests the neglect of which might at times be in the highest degree culpable. But it may be replied that for the expression of an opinion, an act of government is not needed, that our people have the means of expressing their opinions without the instrumentality of its government.

But the indignation of a people is regarded in Europe as little better than the outcry of fanatical anarchists, and as the tyrannies of Europe confound in their misrepresentation, the rational republican with the fanatical altruist, nothing could conduce more to the correction of such prejudice, and to the respectability of republican opinions in Europe, than the protest of a government the best regulated in the world, against unjust acts of tyranny.

Armed intervention as we have seen is not the only kind, nor is it required to support other kinds. But being the extreme of intervention, it should only be used in extreme cases, as an expression of the greatest indignation against the most manifest and universally admitted injustice.

It is often both dangerous and unproductive when the opposing parties firmly believe as they often do that they have justice on their side. This fact is too seldom attended to by the opposers of tyranny. Powerful nations should regard themselves as the executives of international law, but should not act at the same time as judges of the law. And on the other hand in disputed cases should act as judges and not at the same time as the executive.

When however a large majority of enlightened nations like jurors have passed a verdict in favor of those universal sentiments which thus become international law, and against certain acts of outrage and tyranny, then the powerful are bound to act as the executives, in all those cases to which these decisions clearly apply.

Pronouncing a protest is a judgement and an intervention, which may induce other nations to follow our example, and thus the sentiments of the world may become distinctly known, and energetically supported.

At least let us do our part!

2. WAS THE ACT OF BRUTUS IN KILLING CAESAR JUSTIFIABLE?

Historians, when they consider the great abilities of Caesar, the beneficent schemes which he was preparing to execute, and the subsequent course of events, the entire ruin of the republic, agree that his death was an irreparable loss to Rome. Many have been led to consider the murder of Caesar as the most senseless act which the Romans ever committed. If the Romans could have foreseen the consequences and could have viewed them as modern historians do, it no doubt would have been a senseless act. Yet I conceive that the justification of an act depends not on unforeseen consequences excepting so far as they are unforeseen through rashness or want of forethought. That the result of many acts in the life of everyone of us, though carefully inquired into, have turned out to be contrary to our expectations is no more the actual than it is the necessary result of our limited capacities.

If then we can show that Brutus even with the maturest reflection could not have foreseen the actual consequences of his act, we shall have cleared him of the imputation of rashness, and then, if there appears sufficient cause for the murder of his friend, the act is justified.

Nothing but a party spirit can admit the justice of denying honest and upright motives to Brutus on the ground that he belonged to the aristocratical, for, though his party did oppose the agrarian laws and the communication of the Roman franchise to the allies, yet it does not follow that integrity and honesty could not also characterize this party.

We have the testimony of Cicero, one of the most scrupulous and honest men of that age, of the noble character of Brutus. Cicero looked upon him with high hopes for his future greatness and notwithstanding his mistake no one can deny that he excelled in the virtues of his age.

No one, not even the profoundest statesman, can fully comprehend the condition and prospects of a state while they are present as he afterwards sees them when they are past. We always refer to the past on counciling for the present. And so did Brutus. What did he find there?

We read of the noble example of Publius Scipio in destroying (as a private citizen) that plotter against the republic (as Brutus' party thought) Tiberius Graccus. We read the example of Ahala in murdering Spurius Melius; and all those examples of noble patriotism which had saved the republic from ruin. The Gracchi were leaders of the mob and so was Caesar. Noble acts of patriotism had rescued the republic from their hands and so they might from Caesar's. Brutus did not then know that the state had grown, within the shell of republicanism, to be the many-headed hydra of which Caesar's dynasty was the first manifestation: and how could he know? We had no precedents from which to judge that such was its condition: in short to accuse Brutus of rashness is to ascribe to him a keener foresight than usually falls to the lot of man.

In seeking for causes to justify or condemn this act in other respects, we must look to the moral standard of those times, for surely a man's actions can be judged only by those principles of morality in which he has been educated;—not that I mean to say that principles of morality differ essentially in different men and different times, but that they are better defined and understood by some than by others.

If every human action were to be tried before the highest standard of morality, very few would probably be justified.

In our day it would not be a justifiable act for a private citizen to take the safety of the government into his own hands contrary to the course prescribed by law. Yet this was a justifiable act of patriotism, at least according to the tenets of the aristocracy of Rome, when a popular leader threatened to subvert the government.

The charge of ingratitude often brought against Brutus must be examined in the same light: it then becomes the question: which to a Roman conscience appeared the most binding, the obligations which Brutus owed to Caesar or those which he owed to his country.

If we consider the firmness with which Brutus's party stood by the government and the sacred obligations which bound them to maintain it, I think that we can conceive Brutus murdering Caesar not, without gratitude, but, in spite of it. Then will he appear the noble and upright Brutus which his contemporaries have represented him.

C

C. S. Peirce, "Why I Am a Pragmatist"[1]

A persuasive article on pragmatism should begin by showing how some great historic puzzles, such as that about the probabilities of causes, may be solved instantly by the pragmatic method of attacking them, as salt is dissolved in water. It should next proceed to show the nature of mathematical demonstration and of all abstract thought that comes to any settled agreement, chiefly by examples. It should then show on evolutionary and other grounds that intellect is an adaptive character; so much so, that Leibniz denied it of the Deity. Now to what is it adapted, unless to forming habits of action? The next step in the argument should consist of an examination of the processes whereby we gather the meaning of an abstract term. The first of these is attention; and it could be strikingly shown *what glaring differences of sensation we overlook when they have no practical import.* It must be acknowledged that there is a passive attention to sensations apart from any practical considerations. One cannot waive aside a tooth-ache as of no importance. But the sensations thus forced upon us have no intellectual value or import; and the maxim of pragmaticism relates exclusively to this. A critical inquiry into the processes of thought whereby we learn the meaning of such abstract words as 'existence,' 'relation,' 'generalization,' will show that we can only get at their meanings in one or other of two ways; namely, either first, by imagining ourselves to have certain habits of action, or second, by examples of their use calling up into our minds concepts which were already to be called up (or to use an Aristotelian phrase, were in a state of first energy). But if we try accepting the second alternative, the question will arise, How did these ideas ever get into our minds? The answer must be as before: either they had a pragmaticistic origin or they lay hidden in our minds already. In short we are reduced either to saying that the ultimate intellectual purport of a word conforms to the pragmaticistic programme or to saying that abstract ideas are part of our original instinctive nature. For my part, I think *both* are true. But everybody nowadays agrees that all our instincts are pragmaticistic. That is what Darwinism means above all else. Our instincts have a practical purpose. So, then, to reduce a concept to instinct, as the Scotch philosophers very justly do, is really to say that its intellectual import and value lies in general

[1] Unpublished fragment by Charles S. Peirce, at Widener Library, University Archives, Peirce MSS, I B 1, box 1.

modes of action. This argument could be buttressed by others, such as that of my paper entitled 'The Fixation of Belief' (*Pop. Sci. Monthly* for Nov. 1877, vol. xii, p. 1), the whole purpose of which was to prepare the reader for the pragmaticistic method of thinking.

The following brief statement will go too deep to be persuasive. It will simply skim the cream of forty years of thought.

My classification of the sciences purports to be a natural classification such as the taxonomical biologists draw up; and here, less for the sake of the opinion itself than for the illustration it affords of my method of thinking, I wish to express dissent from a certain form of reasoning that is universal among the taxonomists. Namely, they are in the habit of taking two forms that appear to be markedly different, and if they can find specimens to bridge the interval between them with a "series" in which successive numbers show hardly any perceptible difference, they say that those two forms belong to the same natural species and that there is but a single natural class to which all the specimens belong. But I say, on the contrary, that it may be a hard fact of nature that a whole consists of three or more different divisions, and yet it may be a pure matter of arbitrary election to which of those three divisions almost any individual member shall be regarded as belonging. I will prove this by an example. Let us imagine a flat piece of dough, such as would make a cruller or dough-nut if it were

fried; and imagine this piece to have two holes in it, thus Now by moulding the dough, *without joining any parts that are separate or separating any parts that are joined,* we can give the dough an endless variety of shapes, such as these:

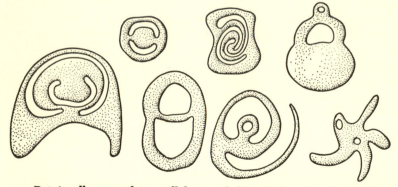

But in all cases, there will be two holes which make the dough to consist of three parts, the part between the holes, the part between one hole and the outside boundary of the dough, and the part between the other hole and the outside boundary of the dough.

D

Peirce's Manuscript on "Hume on Miracles and Laws of Nature"

In a manuscript dated 1901, now at the Smithsonian Institution, Peirce divided Langley's assignment of the topic, "On the 'Laws of Nature,' as understood by Hume's contemporaries and by our own, with special reference to his argument on miracles," into eleven subsidiary questions, as follows:

I. *What is a Law of Nature?* Peirce's answer was: "a predictive [prognostic, foreknowing] generalization of observations," and raised the question, how can the reason of man attain such foreknowledge, unless "the fact that he can so attain proves that there is an energizing reasonableness that shapes phenomena in some sense, and that this same working reasonableness has moulded the reason of man into something like its own image?"

II. *What conception of a Law of Nature was entertained in England in Hume's day, not by those who wrote upon the subject, but by the silent mass of educated men?* Here Peirce pointed to Ockhamist nominalism as the oldest opinion rife in Hume's day, and "which was developed in the first half of the XIVth century, and has had a very strong following in England from that day even to our own [for example, Karl Pearson] without yet betraying any great signs of enfeeblement. This opinion is, that there is but one mode of Being, that of individual objects or facts; and that this is sufficient to explain everything, provided it be borne in mind that among such objects are included *signs,* that among signs there are *general* signs, that is to say, signs applicable each to more than a single object, and that among such general signs are included the different individual *conceptions* of the mind . . . I will simply aver, from having analyzed the whole argument, that the Ockhamists are forced to say of a law of nature that it is a similarity between phenomena, which similarity consists in the fact that somebody *thinks* the phenomena similar. But when they are asked why *future* phenomena conform to the law, they are apt to evade the question as long as they can." Peirce, in this remarkably clear statement of the nominalist position which he thought all modern philosophy was tainted with, attacks three forms of it: (a) the uniformity of nature as an ultimate incomprehensible fact, (b) disembodied revelation, (c) God's will—which Peirce calls an explanation *à la turque* even when used by Descartes, because it "rendered philosophizing so reposeful!"

III. *What conception of laws of nature is entertained today by the generality of educated men?* Against Langley's opinion Peirce thought it is "the same Ockhamistic conception which was commonest in Hume's time; for most men I meet, when they refer to such matters, talk the language of Mill's *Logic.*" This includes Mill's appeal to the uniformity of nature as an "ultimate fact" which Peirce declares, as he did in 1860 in conversation with Chauncey Wright, incompatible with the idea of evolution, an idea which was destined to alter scientific logic and cosmology.

IV. *What is the conception of law entertained today by typical scientific men?* Peirce's answer is prope-positivistic and pragmatistic: "It does not belong to the function of a scientific man to ascertain the metaphysical essence of laws of nature. On the contrary, that task calls for talents widely different from those which he requires. Still, the metaphysician's account of law ought to be in harmony with the practice of the scientific man in discovering the laws; and in the mind of the typical scientific man, untroubled by dabbling with metaphysical theories, there will grow up a notion of law rooted in his own practice."

V. *What was Hume's argument against miracles?* Here Peirce divided Hume's tenth chapter of his *Enquiry Concerning Human Understanding* (1748), "On Miracles," into two parts, the first consisting of the first thirteen paragraphs "which contains the gist of the whole," and the second part consisting of the remaining paragraphs. After condensing the argument of each paragraph in a sentence, Peirce made historical and critical comments on each. He especially scrutinized Hume's remark "a wise man proportions his belief to the evidence," by defining "evidence" much more rigorously than Hume and in the sense of the calculus of probabilities (which Peirce illustrated in a problem). He concluded that Hume erroneously assumed in line with his nominalistic metaphysics that single instances of an induction are *independent* "evidences." Whence Hume was hopelessly wrong in talking of the numerical veracity of a witness, though if "balancing evidences" be interpreted not numerically nor in the frequency sense of probable inference, but psychologically, "then with the proper definitions, there can be no doubt that his [Hume's] method of 'balancing evidences' is profound and excellent, and that his argument does refute all very extraordinary histories." This is the most technical part of Peirce's manuscript, and should be of interest to logical theorists of probability as well as to students of the history of the subject.

VI. *What is the bearing upon this argument of any particular conception of the laws of nature, or of the metaphysical essence of a miracle?* Peirce here indicated that Hume did not have to say that a miracle was impossible or a violation of the laws of nature in the sense of Aquinas' "order of nature," but should have simply stuck to the position of the church fathers that a miracle was a "marvel" or something very unusual in human experience.

VII. *What effect did Hume's argument produce upon the minds of those of Hume's contemporaries who did not publish their opinions?* Peirce's opinion was that "it really only appealed to Ockhamists, very few of whom were not already dead-against miracles."

VIII. *What bearing did those persons who did not express their opinions in print conceive that the metaphysical essence of a law of nature had upon Hume's argument?* If "law" meant "decree of the Almighty," then Hume's argument "could, and did, lead men to argue that what would be exceedingly improbable under ordinary circumstances, might become highly probable when God wished to reveal himself." But such reasoning is too abstruse for most men, and nonsensical to the Ockhamist.

IX. *Does Hume's argument affect the educated men of our day differently; and if so, why?* Peirce remarks here that "very few Christians now rest their faith on the gospel miracles. They believe in the religion because of some personal experince of their own." (Note that William James was at this time preparing to deliver his Gifford Lectures on *The Varieties of Religious Experience.*) The German critics, adds Peirce, have exploited Hume's psychological approach in their historical theories of Christianity. Hume "was really one of the great geniuses in psychology."

X. *Has any change in the usual conception of a law of nature been influential in causing the change of attitude toward Hume's argument?* Peirce's reply ran squarely counter to Langley's original notions which had led to all the correspondence and the various versions of Peirce's paper: "The very same conception of a law of nature which was most widely adopted in Hume's day is certainly the one most widely adopted now . . . It is true that a section of the scientific world has of late been considerably occupied with the essentially provisional character of scientific theories. There is nothing new about this. St. Augustine, in connection with miracles, remarked that unregenerate man can know nothing of the real order of nature, and that a miracle is but a violation of the laws of nature *so far as we know them;* and Bishop Butler, in Hume's day repeated the remark." There is nothing in Ockhamism to make miracles *impossible,* for positive experience as the sole support of belief logically makes miracles *improbable.* Peirce concludes this part with another implied reference to his own metaphysical evolutionism: "So far as there are, in our times, evolutionary tendencies, which already begin to cause us to regard laws from a different point of view, there is the germ of ideas destined to destroy Ockhamism, and with it the argument of Hume against miracles."

XI. *What are the real merits of Hume's argument?* Against Hume's empiricism, Peirce argued that "every item of science comes originally from conjecture, which has only been pruned down by experience." An inadequate Ockhamistic logic of hypothesis has produced ridiculous results by modern German critics of ancient history, "with whom it is a constant practice to deny the testimony of all the witnesses and to set down what seems likely in a German University town in the place of history . . . The same principle has been practically applied on the continent of Europe to judicial evidence; and if anybody is shocked by the hideous wrongs perpetrated by our own courts and district attorney's offices, he can find some balm for his wounded Americanism in studying the results of the method in question in continental 'justice.' " [1]

[1] Major A. Dreyfus' *Five Years of My Life* appeared in 1901.

E

Arthur O. Lovejoy, "A Note on Peirce's Evolutionism"

The interconnection of Peirce's ideas here[1] is curious, and some remark upon it may perhaps be worthwhile. He seems to argue thus: 1. A cosmological or cosmogonic theory is more deserving of acceptance if it satisfies the "religious instinct." 2. The religious instinct leads us to believe in a God as Creator of the "universe of nature." 3. It therefore leads us to believe that the created world has a character which is worthy of its Creator. 4. The "theory of objective chance" is in accord with these two beliefs. 5. Therefore, that theory is deserving of acceptance on religious as well as other grounds; it satisfies the "religious instinct."

Why does it? one asks. Peirce's answer is: Because it represents the created world as originally a pure chaos, a world without laws (i.e., without fixed habits), with "nothing rational in it"—but out of which, purely in consequence of the "law" of arithmetical probabilities, rationality gradually and increasingly and *necessarily* emerged. But *why* was it more "worthy of the Creator" to create a chaos than to give the world *some* habits—a certain amount, at least, of rationality—from the outset? Would not the supposition that he did so be *more* satisfying to the "religious instinct"? And does the "theory of objective chance" give any *reason* for postulating a Creator? A chaos is a world of chance—one in which no one kind of existence, and no one, and no limited number, of "ways of happening," is more probable than another, within the limits (if any) set by the whole number of possible ways of happening. (Does Peirce postulate such limits?) And even though, according to the "theorem of probability," order, laws, particular and fixed ways of happening, must increasingly preponderate over the original chaos, this is a purely logical and, so to say, automatic necessity.

But the conception of a Creator-God has usually been supposed to explain why the creation has one character rather than another, and it is at least assumed by Peirce that there ought to be a correspondence between its character and that of its Creator. But a) if its initial character was that of complete promiscuous diversity, and variability without any laws of variation, it could hardly be said to have any "character" as a whole—unless this absence of continuing and determinate character be itself called a

---

[1] The quotation from Peirce, referred to by Professor Lovejoy, occurs as the penultimate paragraph of Chapter IV above.

"character." In the chaos, everything that could be, or could happen, sooner or later was, or did happen—in entire independence of everything else that was, or happened; for if one variation had been correlated with any other, there would have been a "law." b) If, however, you argue back from the assumed original "character" of the world to that of the Creator, he also would seem to have no particular character, in the sense of preferring one kind of world rather than another—except that the one thing which you would have to say that he did *not* prefer was "rationality" or "order." Whatever else be said of this, it hardly seems religiously edifying. c) Moreover, Peirce's sole reason for believing in his original chaos is that, while regularity needs to be "explained," irregularity does not; it is apparently for him self-explanatory. No theory, therefore, would seem less to *require* the hypothesis of a Creator.

Aside from this theological aspect of Peirce's cosmological theory: *why* does he assume that a chaos would necessarily become increasingly less chaotic? The question is pertinent to the quotation from Peirce on page 81—[2] which is the crucial argument both with respect to the analogy between Darwinism and Peirce's cosmological theory, and to the justification of that theory. The assumption necessary to permit the inference that a chaotic world would, of necessity, become gradually more "reasonable" is that (in the situation postulated) there are "large numbers of objects having a tendency to retain certain characters unaltered," i.e., there are assumed to be some "objects" which already have the habit of remaining chiefly, though not wholly, as they were. This corresponds to the factor of heredity in the Darwinian theory; in spite of the slight chance deviations in the offspring, the latter in the main have the same characters as their parents. But if this tendency to "stay put" is the Darwinian counterpart of Peirce's "cosmic reasonableness," it seems difficult to understand why there should be any "evolution," any increase of reasonableness. So far as this factor operated, it would tend to produce fixity of characters or of ways of behaving, whether in organisms or in "objects" in general; things would continue to be what they were in the beginning.

The *evolutionary* factor, the one making for the development of new species, in the Darwinian theory, consists initially in the chance variations. These do not, in any given generation, preponderate over the fixity-factor (heredity), but in the course of many generations certain kinds of variations tend to *become* fixed, and the accumulation of a number of these produces changes in specific characters, so that the descendants of a given kind of

[2] "This Darwinian principle is plainly capable of great generalization. Wherever there are large numbers of objects having a tendency to retain certain characters unaltered, this tendency, however, not being absolute but giving room for chance variations . . . there will be a gradual tendency to change in directions of departure from them."

organism may differ from it widely, and even lose some of the characters of their remote ancestors—while other variations are not perpetuated and speedily disappear. But why? Because, says Darwin, of the selective action of the environment; the adaptive characters are conserved and become fixed characters (so long as the environment remains relatively constant), the nonadaptive ones are eliminated.

Now this means that the theory of natural selection, as an explanation of organic evolution, is *not* simply a "theorem of probability." If you knew the range of possible variations, and assumed complete randomness, you might, by applying the calculus of probabilities, predict how many kinds of chance variations would appear among the offspring of a certain number of pairs of parent-organisms of a given species in a given time; the probability would be that, the more numerous the births and the longer the time, the more nearly the numbers of the several kinds of variations would approach equality. But you could not predict, by any calculus of probabilities, that certain of the variations would be *selected* for survival and augmentation, and certain other kinds be eliminated. This requires the assumption of an empirical or factual causal determinant, a quasi-mechanical factor. You must assume that some of the dice (i.e., variations) are loaded (i.e., adaptive), others not. Now, is there any analogy between this and Pierce's assumption of the evolution of "concrete reasonableness" in "nature" as a whole? There does not seem to be in the latter any external factor playing the rôle of the selective action of the environment in Darwin; and to this extent, at least, the analogy between Darwin's evolutionary biology and Peirce's scheme of cosmical evolution breaks down.

Peirce does deduce the evolution of "concrete reasonableness" from a theorem of probability alone—or professes to; Darwin does not deduce the evolution of new species from a theorem of probability alone. How does Peirce do so? The question might be concretely put thus: Suppose (as Peirce does) an early—though not strictly primordial—state of the physical universe. Suppose that in this state there are (to keep the numbers small) 100 "objects," of which 30, to be called Class A, "have a tendency to retain certain characters unaltered," though with "some room for chance variations" (which objects, in other words, are *already* preponderantly "reasonable"), while 70 of the objects, called Class B, have no such tendency, but alter their characters to an indefinite extent and in a perfectly lawless manner. Given these suppositions, Peirce seems to argue, Class A will in the course of time become still more numerous, relatively to Class B. Why? Peirce's answer apparently is: Because, since the objects in Class B vary their characters promiscuously in all possible ways, and since one possible "character" is that of "retaining" *other* "characters unaltered," *some* objects in Class B will sooner or later, in their random variations, acquire *this* character (call it R)—of *not* varying, or at least of not varying so much and so promiscuously. Consequently, the number of objects of Class A will be increased by accidental additions from Class B, but not *vice-versa*. Thus in the long run Class A will become more and Class B less numerous, and in the process of the ages things will lose all their original "spontaneity" and mutual independence, and the world will become "an absolutely perfect, rational and symmetrical system." (There seems, incidentally, to be a curious wavering in Peirce's valuations; he sometimes writes as if "spontaneity," "freedom," or "irrationality" were a good thing, at other times as if the best possible state of things would be one from which it was completely excluded.)

The "trick" of the argument, if I may put it so, consists in including among the possible random variations a tendency to depart from random variation, and in assuming that, in a finite time, this variation, R, will inevitably appear again and again. There seems to be no counterpart to this crucial assumption in the Darwinian theory. And it is not clear to me that it is a permissible assumption. For in the chaos the number of possible random variations must be either infinite or limited. If it is infinite, there is no assurance that variation R would turn up in less than an infinite time. And if the number is limited—i.e., if some conceivable possibilities of variation are excluded—there is no assurance that variation R would be among those included. It *might*, upon such an assumption, be one of the arbitrarily favored possibilities; but is there any reason given by Peirce for supposing that it will or must be? If not, his attempt to deduce the necessity of the emergence of regularity from the conception of "absolute chance" appears to lack cogency—even if there were no other difficulties in his "tychistic" cosmogony, as I think there are.

F

Biographical Notes on Nicholas St. John Green (1835–1876)

Nicholas St. John Green was born at Dover, New Hampshire on March 30, 1830, and died in Cambridge, Massachusetts, September 8, 1876. Chauncey Wright, also born in 1830, knew Green while they were preparing for Harvard as students of the Rev. Rufus Ellis at Northampton, birthplace of Wright, J. B. Thayer, and W. E. Gurney. Green was present with the novelist Henry James at Wright's deathbed in September 1875. Wright was for the last years of his life Green's most intimate friend and a frequent visitor at his house at 8 Story Street in the block adjoining Wright's room on Brattle Street. John Fiske was also a close friend, and on Sunday mornings Green—who was not a churchgoer—often went with his son, Frederick, to visit the portly, pipe-smoking, beer-drinking historian and cosmic evolutionist in his home on Berkeley Street. Other Sunday calls were made to Dean Gurney and to Henry W. Paine, a leader of the Boston bar. Green's father was the Rev. James D. Green, a Harvard graduate of the class of 1817, who changed from the Congregational to the Unitarian ministry, preaching at Lynn and East Cambridge, and then retired from the ministry to be mayor of Cambridge in 1847, 1853, 1860, and 1861. Nicholas Green's mother was a daughter of Judge Durell of Dover, New Hampshire who at one time was congressman for that district. Her brother was Edward Durell, a mayor of New Orleans, appointed by President Lincoln to be a district judge in Louisiana. She was a granddaughter of John Wentworth, a signer of the Articles of Confederation, and own cousin to "Long" John Wentworth, several times mayor of Chicago, and congressman. Thus Green had many relatives active in the judicial and political history of the country.

When Green was graduated at Harvard College in the class of 1851, with James Bradley Thayer and Christopher Columbus Langdell, he was much interested in medicine and started as a student in the medical school, but gave it up because the sight of blood made him faint and sick. So he studied law instead, under Story and Washburn. One of his closest friends at college was Joseph H. Choate of the class of 1852, who became a leader of the New York bar. In Boston, Green became a junior partner of the spectacular criminal lawyer, Civil War general, and politician, Benjamin

Franklin Butler (1818–1893). Their practice was chiefly in criminal and personal injury cases, in which Green found much of human interest. He liked Butler, though not perhaps without some reserves. They were alike in their sympathy for the unfortunate and in their dislike of the self-styled aristocrats of Beacon Hill whose pretensions Butler always took delight in puncturing. The partnership ended when the Civil War broke out and Butler became a major-general. Green soon enlisted also and served as paymaster under Butler at Norfolk. Later he resumed practice in Boston by himself, and continued in it until his death.

After the publication of a distinguished scholarly article on "Proximate and Remote Cause" in the *American Law Review*, and probably in part because of it, Green was appointed lecturer in the Harvard Law School in 1870. The appointment was renewed for the next two years. Meanwhile, his classmate, C. C. Langdell, credited with being the iconoclastic originator of the case method of teaching law, was made dean of the law school. Though Langdell is called the founder of the "case system," his method of teaching was substantially different from that used afterwards at the school, and it is the latter that is connoted by the phrase today. The previous practice at the Harvard Law School, as at law schools in general, had been to have students read textbooks on law and then attend lectures on the topics in these treatises. This is the same sort of thing done today so frequently in college courses in history, philosophy, and other subjects. Langdell substituted the reading of reports of selected cases for the reading of treatises on law, and his lectures consisted of comments on these cases. However, under the case system developed by his successor Ames and others, the students take a more active part in class, stating the cases, the facts, the decisions and reasons for them, and criticizing them under the control of the instructor. Green did not like Langdell's method of teaching, and when some Boston lawyers induced Boston University to start its own law school in 1873 he was offered a position there. On mentioning the offer to President Eliot, Green was told that he could not teach in both law schools. Since the Harvard salary was the larger, perhaps the President thought that the matter was ended. It was decisive, but in the opposite way. Green, wishing to keep his independence, chose the position with the lower salary. At the Boston University Law School during 1874–75 Green was asked to perform the executive duties of Dean Hillard during the dean's illness. He also lectured on Kent's *Commentaries* and on Torts, and continued to act as dean until his death.

Prior to his leaving Harvard, Green had taught not only in the law school but in the college, where he gave courses in philosophy (logic, metaphysics, and psychology) and political economy. It appears that he wrote some book reviews for the *Nation*, but they have not been identified. His son also reports that he has lost the manuscript of his father's review of McCosh's *Logic,* and that the review was never printed, presumably because it was severely critical, ending with "We do not say this is the worst possible book on Logic, but it is the worst we have ever seen." Unlike so many university teachers, Green was evidently not daunted in his thinking by the authority of college presidents. Likewise, in his legal teaching he had contempt for the maxim *stare decisis.* The obituary notices in the law journals and Boston newspapers give us a glimpse of the man as an exceptional lawyer, teacher, and independent thinker:

"Mr. Green was a strong character. He was full of earnest endeavor to strengthen the school, and fond of his students. His weakness, if he had any as an instructor, was his contempt for the maxim *stare decisis.* He loved to attack adjudications. He had a great fund of good nature, of which the students often availed themselves during his lectures by questions which were not always relevant to the point at issue and which he always received pleasantly and in fact seemed to enjoy." [1]

"Mr. Green graduated from Harvard College twenty five years ago, and from the Dane Law School ten years later [1861]. Soon after his admission to the bar, he entered into partnership with General Butler, continuing with him five or six years, gaining a high reputation and an extensive practice. At the close of the war, Mr. Green was appointed an instructor in mental philosophy, in Harvard College, which position he held with distinction until his appointment as lecturer in the Dane Law School. The latter post he occupied for several years and in it achieved great success. Three years ago he refused the lectureship to accept a similar position in the Law School of Boston University. His honorable and successful connection with this institution is widely known. For two years prior to his death, he was Dean of the department, and to his ability and management the success of the Boston Law School is very largely due. As a lawyer Mr. Green had few equals. He argued some of the most intricate cases which are reported in the Massachusetts Reports. As a legal critic his abilities were of the highest order. In his death, there is a vacancy left which will not soon be filled." [2]

[1] Obituary notice of Nicholas St. John Green, *The Green Bag,* I, 54.

[2] *Central Law Journal,* III, 603 (Sept. 15, 1876). *Albany Law Journal,* XIV, 236 (Sept. 22, 1876). Nicholas St. John Green, Dean of the Boston Law School, died at his home in Cambridge on September 8, 1876. "He had been suffering for some time, and to relieve himself, took an overdose of laudanum which resulted in his death." The following verses by a George Sennott were published in a Boston newspaper soon after Green's death, and later in Littell's *Living Age:*

Nicholas St. John Green

Dear Friend! The ancient elegiac strain
For Death, was death itself and dark despair,
Each word a sob; the vain lament in vain
Fell on the careless air.

Extracts from notices published in the Boston newspapers follow:

". . . for one of his years, but few have done so much and so well. He had a fine legal mind, and although more especially attracted to special branches of the law, he was well grounded in all its essential departments . . . He was a man of learning and accomplishments in other branches than the law. Metaphysics were his study and his delight. He had a warm heart, and was ready to make any sacrifice to serve a friend."

"As a lecturer, few have equalled him in the clear, graphic and impressive manner in which he placed his subject before the students. He was not only thoroughly conversant with his subject himself, but had the happy and rare faculty of imparting that knowledge with a directness, simplicity and interest that the ordinary mind that gave him attention could not fail to understand and appreciate. Often with commonplace illustrations, by imaginary incidents and events, the point that he wished to fix on the mind and memory of the student was so strongly set forth that no one could fail to see its scope and comprehend its application. With students he was always exceptionally popular, and by his courtesy, kindness, patience and geniality, gained not only respectful listeners, but won many warm personal friends and ardent admirers. He was no man's mouthpiece, and could give a reason for the faith that was in him. The number, the oppressive dignity of judges, or their apparent opposition against him never awed him into acquiescence of a principle or a decision that he believed wrong or not fully affirmed in the premises and reasoning on which they were claimed to be based. As a man we hardly think he was in complete sympathy with the peculiar phases of life about him, to which we must always conform to a certain extent to draw us the popular applause of the community in which we live . . . He could not cut and trim his ideas to any formal style, however widely he saw it adopted by his fellow citizens. His nature, like his stature, was erect, independent and unfettered. But our subject is greater than our pen."

Better our teachings—though they teach as well
How deathless atoms in eternal flow
Compose these mortal bodies where we dwell,
And all things high and low.

Can senseless atoms live—forever live
And that which animates them ever die?
Can we together brought without our leave,
Then forced apart to fly,

Be worse for immortality? Our loss
Cannot be lasting while He lasts, to tell
What glory shines behind his bitter cross
Who doeth all things well.

G

Biographical Notes on Joseph B. Warner (1848–1923)

A lawyer's professional life is by definition one of actions either in the courts or, more frequently, in settlements out of court, so that it is not surprising that the historian of American pragmatism will inquire into the history and philosophical ideas of lawyers, even when, as in the case of Joseph B. Warner, they are hidden in a brief record of biography and publication. If Peirce had not mentioned him as a member of the Metaphysical Club,[1] Warner would not figure at all in the history of American philosophic thought, but he may be viewed as reflecting and criticizing some of the evolutionary and pragmatic ideas of Wright as his teacher, and of Green and Holmes, at work in the practical business of American law. Let us look at the little we know of his life and scant writings.

The *Boston Transcript* of January 2, 1923 contains the following death notice:

"Joseph Bangs Warner, prominent until a few years ago as a Boston lawyer, died last night at the family home, 23 Brimmer Street, this city. Mr. Warner was born in the old Fort Hill square section of Boston, August 5, 1848, and was the son of Caleb Henry Warner, president of one of the old-time banks, and Elizabeth (Bangs) Warner, and was a descendant of William Warner of Ipswich, who settled in that town in 1636. He studied at Harvard and received his A.B. degree in 1869 and his A.M. in 1872. He studied law in the office of William G. Russell of Boston and entering the Harvard Law School received his LL.B.[2] and began the practice of law in

---

[1] *Collected Papers of Charles Sanders Peirce*, edited by C. Hartshorne and P. Weiss, 6 vols. (Cambridge, Mass., 1934–1936), 5.12.

[2] Joseph B. Warner while at the Harvard Law School was a student of C. C. Langdell, Dane Professor of Law. "Young Warner, a keen logician (and one of the first converts to the new ideas), cornered him [Langdell] squarely one day, amidst a hurricane of derisive clapping and stamping. Would it be believed, 'the old crank' went back to the same point the next time and worked it all out to a different conclusion. Most of the class would see nothing in this system but mental confusion and social humiliation. They began to drop away fast." "It was Ames who fully developed the Socratic method of teaching: Langdell never encouraged much discussion in class, and in later life was so brimful of his subject that he confined himself to straight lecturing" (Samuel F. Batchelder, *Bits of Harvard History* (Cambridge, Mass., 1924), pp. 316, 318n.1).

Boston being a member of the firm of Warner, Warner & Stackpole, which firm today is known as Warner, Stackpole & Bradlee. Mr. Warner continued to practice until 1916.

"Mr. Warner was a trustee of Radcliffe College and at one time had been a trustee of Simmons College. He also had been an instructor in history at Harvard during the years 1872 and 1873, and was a lecturer in the Harvard Law School in 1886 and 1887. At one time he belonged to several organizations, but had relinquished all associations some time before his last illness which began seven years ago. In his legal practice Mr. Warner was an active factor in bringing about the big Boston & Maine merger a number of years ago.

"In 1876, Mr. Warner married at Cambridge, Margaret Woodbury Storer. His two sons are Roger Sherman Warner, and Langdon Warner, and there is a brother Henry E. Warner of Boston."

At Harvard, he was a student in Chauncey Wright's small course of university lectures on psychology in 1870, in which Bain's work was used as a text. Warner testified as a "friend" in his letter of recollections in J. B. Thayer's *Letters of Chauncey Wright,* that the lectures were not successful because Wright had the habit of talking "in a monotonous way, without emphasis, and they failed to arouse interest . . . In looking over the notes which I took at the time, I find that he began at the very beginning, and ventured to expect no philosophical training, and hardly even a knowledge of philosophical terms, in his hearers. In this sort of work, he had had, I suppose, no experience. A poorer man might have done it better. The class did not aid him much by discussion,—a thing for which he expressed regret to me in private. But his monotonous fluency seemed, to those of the class who did not know him, to forbid interruption. He showed the utmost patience; but I do not think that he quite knew how to approach the class . . . He did not talk over our heads; but he failed to interest us. You may think it strange, for you have undoubtedly seen him, as I often have, interest and instruct children and persons without special training. But what he could do in conversation, stimulated by questions, and himself interested in the effect of his teaching or talking, he could not (or at least did not) do in these lectures, where he had not one point to expound, but a system to cover, and that to a knot of persons who made little response but devout scribbling in their notebooks." [3]

This letter of Warner may throw some light not only on Wright as a teacher but on Warner's early empirical and psychological mode of evaluating educational method in terms of practical experience on the part of the

[3] *The Letters of Chauncey Wright,* edited by J. B. Thayer (Cambridge, Mass., 1878), pp. 213–14.

teacher in techniques of arousing the interest of his students. Like John Stuart Mill and Chauncey Wright, Warner was deeply interested in higher education for women, and took an active part in founding Simmons College and Radcliffe College.

The next piece of published writing by Warner is his review of Holmes's *Common Law* in 1881.[4] Warner had helped Holmes in editing Kent's *Commentaries on American Law* in 1873. Hence, Warner's review was based on an intimate acquaintance with Holmes's background and historical approach to the law. Something of Wright's cautious attitude to unrestrained evolutionism, is reflected in Warner's comment on Holmes's historical method: "The great danger in this method is that the theory account for too much." Holmes's articles in the *American Law Review* sometimes reveal an "overrefinement of elaborate historical explanations of ideas which have their sufficient source in the constant traits of human nature." [5] However, Warner finds Holmes, when he is not looking too deep into historical customs as "survivals" in the law, remains on the ground of common sense allowing spontaneity and the exigencies of life as "the great controlling agencies, however tenaciously old ideas may cling to the rules of law." Holmes's *Common Law* shows how the living power of man interferes in history to make things as he wants them, and this power of interference is an essential part of the evolutionary process in the human world. But again, the ghost of Wright's critique of the excesses of evolutionism appears in Warner's ensuing remark that Holmes goes further than the evidence warrants when he emphasizes so much the constant adaptation of law to the needs of the community. Warner does not use our modern sociologist's phrase "cultural lag," but as a practical lawyer he is aware that the law moves slowly. Though "considerations of convenience and policy must in the end prevail, it cannot be denied that the dead hand of the old law lies with no light pressure upon the courts." [6] Holmes's book makes it probable that the adaptability of the law is increasing, especially through the legislative office of the courts. The law is becoming to be understood as what the judges say it is.

There is, in Warner, the same reaction against German idealistic ethics of an intuitive or metaphysical sort (Kant, Hegel), as we find in Wright and Green. He lauds Holmes's utilitarian and evolutionary theory of the "external standard" of judging right and wrong in terms of objective, historical consequences for society rather than by reference to unverifiable intent on the part of the defendant. We have thus made some progress in the law from its earliest origins in private vengeance or family feuds.

---

[4] *American Law Review* (May 1881), p. 332.
[5] *American Law Review* (May 1881), p. 333.
[6] *Ibid.*

Holmes "manifests the tendency in modern ethics toward objective grounds, and away from intuitive and metaphysical theories of morals." [7]

The application of the external standards of conduct entails certain logical and social features. Logically, the law raises hypothetical questions of fact: "Would a man of average prudence have done so?" Warner does not tell us how to find out what a man of average prudence does, but he does suggest that direct analysis needs to be supplemented with the results of historical research, such as Holmes's *Common Law* provides.[8] In contrast with Langdell's mode of treatment of such topics as consideration in the law of contracts, Holmes's approach is closer to the life of the law. Clearly Holmes's work is by no means solely historical in purpose for a good part of it is devoted to analyses of existing laws, for example, of possession, contracts, ownership, and real estate covenants. "The study of the genesis of an idea adds an entirely new interest, and gives an almost inexplicable ease of comprehension when the direct analysis of the idea is undertaken." [9]

Thus Warner strives to maintain a practical man's equilibrium between the excesses of historical evolutionism in which an overoptimistic belief in progress prevails and the static conception of legal logic in which the judge simply applies abstract rules to particular cases. Green, Warner, and Holmes found Langdell's "case method" a mechanical travesty on the historical life of the law. A truly inductive method should trace the development of the law in response to changing conditions instead of simply using the cases of law as illustrating abstract rules which do not in fact operate in the historical making of judicial decisions.

Ten years after this review of Holmes, Warner was engaged by a distinguished philosopher at Harvard, Josiah Royce, to protect his interests in a possible charge of libel raised by the "radical" theologian, Francis Ellingwood Abbot. It was an unpleasant episode in the philosophical life of academic Cambridge. Abbot had published the lectures he had given in an advanced course in the philosophy department during Royce's leave of absence (1888), under the title *The Way Out of Agnosticism or the Philosophy of Free Religion*.[10] Royce in a more than caustic review in the *International Journal of Ethics*, of which he was co-editor with Felix Adler, ventured "to speak plainly, by way of professional warning to the liberal-minded public concerning Dr. Abbot's philosophical pretensions. And my warning takes the form of saying that if people are to think in this confused way, unconsciously borrowing from a great speculator like Hegel, and then depriving the borrowed conception of the peculiar subtlety of statement that made it useful in its place—and if we readers are for our part to accept such

---

[7] *American Law Review* (May 1881), p. 336.
[8] *American Law Review* (May 1881), p. 337.
[9] *Ibid.*   [10] Second edition, Boston, 1890.

scholasticism as is found in Dr. Abbot's concluding sections as at all resembling philosophy—then it were far better for the world that no reflective thinking whatever should be done." [11] Royce ended his review by saying that he "showed no mercy and asked for none." Abbot in an open letter to the overseers of Harvard asked whether in the light of such a malicious and libelous aspersion on his character, Professor Royce ought not be held accountable. Abbot had also written a reply which Adler held up pending the preparation of a rejoinder by Royce. Charles S. Peirce then appeared in an open letter to the *Nation*[12] defending Abbot's reputation as "a profound student and a highly original philosopher, some of whose results are substantive additions to the treasury of thought . . . Professor Royce's warning is an unwarranted aspersion . . . It is quite impossible not to suppose that Professor Royce conceived it was his duty thus to destroy Dr. Abbot's reputation, and with that the happiness of his life." In withholding Abbot's reply, Felix Adler, the editor-in-chief, "does not appear as very strong in the practical department of ethics. Afterwards Professor Royce, through his lawyer [Joseph B. Warner] threatened Dr. Abbot with legal proceedings if he published his reply at all."

William James, friend and colleague of Royce, came to his assistance, by criticizing Peirce's statements as not founded on sufficient knowledge of the facts. Warner then published a letter in the *Nation*[13] on behalf of his client, Royce, to the effect that Abbot's reply was withheld because Abbot had refused to omit a personal attack. Finally, Abbot had the *Nation*[14] print part of his unprinted reply to confute Warner's charge of being personal.

Twelve years later, Abbot voluntarily took his life by means of poison while visiting his wife's grave. In his pocket was a brief letter stating that he had completed his life work—a work which had its inception in his college days.[15] That work was a positivistic religion of evolutionism which led Abbot to leave the Unitarian ministry in 1863 when its freedom was restricted by the Cambridge right-wing Unitarians, to found the Free Religious Association, to edit *The Index*, and to write his many books on behalf of a scientific theism based on evolution.[16] Peirce had regarded him as one of the strongest thinkers he had ever known, and there is some affinity

[11] *International Journal of Ethics*, vol. I (1890).

[12] C. S. Peirce, "Abbot Against Royce," *Nation*, LIII, 372 (Nov. 12, 1891).

[13] *Nation*, LIII, 408 (Nov. 29, 1891).

[14] *Nation*, LIII, 426 (Dec. 3, 1891).

[15] *Boston Transcript* (Saturday, Oct. 24, 1903). Cf. "The Late Dr. F. E. Abbot" in the same newspaper (Thursday, Nov. 19, 1903) by Edward C. Towne of Albany, New York. The newspaper clippings are in the New York Public Library.

[16] See *Dictionary of American Biography*, article on "Abbot, Francis Ellingwood," by Francis A. Christie. See also, Chapter II, note 14.

between their religious philosophies of evolution interpreted in terms of Scotistic realism.

Was it, among other things, the Royce-Abbot affair that was in Warner's mind when in 1896 he addressed the American Bar Association on "The Responsibilities of the Lawyer" in relation to his client? Warner argued that the lawyer who has to make his living defending the cause of almost any client that comes to him and is at the same time an officer of the court defending justice is in a paradoxical situation. Unless the practicing lawyer maintains *some* detachment or neutrality towards the ethical merits of his client's case or refuses to take it, to the extent that counsel identifies himself with his client's cause, to that extent he is responsible or answerable to public opinion. The professional obligations of a lawyer are not quite the same as those of a public-minded citizen who is generally more free to take sides than the opposing lawyers in litigation. He is called upon to make the best of his client's case in accordance with the rules of conduct which the courts enforce. These rules created in part by the legislature and in part by the handling of the courts "may or may not correspond with the moral demands of the individual observer; it may work poorly in particular cases; it is quite possible that it may permit, in this or that combination of circumstances, a result shocking to the moral sense. But this is the system adopted for the business . . . and such a system the lawyer is called upon to administer. He is not to be taken to task, generally, for the results it may produce . . . nor is his conscience usually to be consulted in the matter. The law, so far as it goes, is his conscience here. In invoking that he is not immoral, he is non-moral—administering a system artificial and positive, for which he is not responsible." [17]

This practical principle of non-moral professional detachment is another instance of that scientific neutrality to ethics which is so central to Wright's thought and which appears in Holmes's amoralism. The law is an anthropological document offering evidence in its historical development of man's tribal passions and powerful prejudices, but nonetheless slowly responsive to the needs of men in the clash of their interests. The lawyer is a hired partisan in the combat, but he has to develop an attitude of impartiality that will enable him to fight effectively in cases of opposite character at different times. Warner believed there is even a sense in which because of his impartiality the lawyer develops a moral sense more vividly than men in other walks of life, for it is with the study and analysis of conduct with which lawyers are always busy—judging men, weighing motives, discriminating alternative consequences in the most difficult of human situations, men fighting against one another for all they can get. Our laws and

[17] *American Bar Association Reports*, XIX, 319–342 (1896). The passage quoted is on page 333.

law courts know no easy solution: "Justice does not descend, in answer to our prayers, pure and undefiled from heaven. It is struck out, with pain and sweat and conflict, in the private disputes of men. Our system is not devised primarily to discover truth, nor is a lawyer chiefly a searcher after truth." [18]

There is the drama of history and evolution in the law because of the conflicts of human emotions that break through in a court trial to express "the visible struggle, old as the world, of all the passions of anger, hate, greed, and avarice, less wild than of old, but still full of their inherited spirit, and now forced into an arena which excepting war itself, is left as the only battlefield for these irrepressible fighting instincts of the race." Warner thus shows the effect of the evolutionary controversies and social Darwinism of the late nineteenth century; for example, he attributes the ruthlessness of some lawyers to "traits which yet survive in the human animal." [19]

The realism of Warner's account of the court as an arena is not meant to imply the abandonment of ideas of public service. It is a grim pragmatic reminder of the tough business of effecting compromises that will keep men working together. "For the advancement of high purposes which are not to end in dreams or resolutions but in action, the one appointed way is to go down among men and work. This way lies open in the practice of law." [20]

The Memorial of the Bar [1923] gives us a glimpse of Warner's mind and character: "His intellectual endowment was of the highest order. His opportunities for study and improvement were unusual. By tireless industry and ceaseless study he brought to well nigh perfection his mental birthright.

"He never studied or practised the arts of the orator, but he was a persuasive and compelling speaker with a rich and ample vocabulary. He argued logically and conclusively, and there was transparent integrity in everything he said or did. He had an abiding common sense and was a master of concise and orderly statement. His temperament was equable, and his manner courteous. He was always the scholar, never the pedant." [21]

As a member of the Saturday Club, he received the following tribute by Moorefield Storey: "Warner was interested as a citizen in bettering the administration of public affairs, as a lawyer in the reform of legal procedures, and he gave much time to both . . . He preferred to give questions a

[18] *American Bar Association Reports*, XIX, 341.
[19] *Ibid.*
[20] *American Bar Association Reports*, XIX, 342.
[21] Quoted by Moorefield Storey in the *Later Years of the Saturday Club, 1870–1920*, edited by Mark A. DeWolfe Howe (Boston and New York, 1927), p. 298.

longer consideration than often seemed to me necessary. He was not impetu-
ous or prompt in action, but his mind was critical and judicial. As an advocate, however, he was tenacious of his views and masterly in presenting
them to the Court."

On January 4, 1916, he was suddenly stricken so that nothing was left
of his life but seven tragic years of suffering during which he was unable
to speak intelligently or to write more than a few words. "Reasoning as he
had known it, was no longer possible, nor premeditated action. This he
knew well enough. But there was no beating against the bars. It was as
if the mind had abdicated and God had taken command. What was left was
what had been innate—courtesy, considerateness, generosity, courage. He
succeeded almost in robbing the thing of its tragedy." [22]

[22] *Saturday Club,* p. 299.

NOTES

I. THE BACKGROUND OF SCIENCE AND NATURAL THEOLOGY

1. Arthur O. Lovejoy has made the most notable studies of pre-Darwinian scientific, philosophical, and literary writings which helped prepare the thinkers of the 1860's and 1870's (when the founders of pragmatism appeared on the scene) to be so favorably responsive to Darwin's ideas. Among the early articles of Professor Lovejoy which appeared in the *Popular Science Monthly* are "Some Eighteenth Century Evolutionists," LXV, 238–251 (1904); "The Argument for Organic Evolution Before 'The Origin of Species,'" LXXV, 499–514, 537–549 (1909); "Kant and Evolution," LXXVII, 538–553 (1910), and LXXVIII, 36–51 (1911). These articles and *The Great Chain of Being, A Study of the History of an Idea,* The William James Lectures delivered at Harvard University, 1933 (Cambridge, Mass., 1936), especially chapter ix, "The Temporalizing of the Chain of Being," deal with the evolutionistic ideas of Maupertuis, Buffon, Bonnet, Geoffrey St. Hilaire, Lord Monboddo, Goldsmith, Erasmus Darwin, Tennyson, Emerson, Herder, Schelling, Schopenhauer, Lamarck, Robert Chambers, Herbert Spencer, and A. R. Wallace. Darwin, in the third and subsequent editions of his *Origin of Species,* mentions about thirty-four of his precursors between 1794 and 1859. On the American precursor of Darwin, William Charles Wells, see R. H. Shryock's article in the *Dictionary of American Biography;* his "The Strange Case of Wells' Theory of Natural Selection: Some Comments on the Dissemination of Scientific Ideas," in *Essays in Honor of George Sarton* (New York, 1947); and his *The Development of Modern Medicine* (New York, 1947), p. 133. On evolutionism and on pragmatism in American philosophy, see H. W. Schneider, *A History of American Philosophy* (New York, 1946), chapters vi and viii; also, M. H. Fisch, "Evolution in American Philosophy," *Philosophical Review,* LVI, 357–373 (July 1947).

2. *Collected Papers of Charles Sanders Peirce,* 6 vols., edited by Charles Hartshorne and Paul Weiss (Cambridge, Mass.: Harvard University Press, 1931–1935), vol. VI, paragraph 297 (designated by the editors as 6.297). Cf. 5.364. Hereafter, *C.P.* will refer to this edition.

3. Muriel Rukeyser, *Willard Gibbs* (New York, 1942). See also, C. S. Peirce's review (unsigned) of *The Scientific Papers of J. Willard Gibbs,* in *Nation,* LXXXIV, 92 (Jan. 17, 1907): "Gibbs advanced science the world over more than it has ever been given to any other American researcher to do."

4. John T. Merz, *History of European Thought in the Nineteenth Century* (Edinburgh and London, 1907), I, 120n.1. See also, Clerk Maxwell, "Illustrations of the Dynamical Theory of Gases," *Philosophical Magazine,* vol. IV (1860), and *The Scientific Papers of J. C. Maxwell,* edited by W. D. Niven, 2 vols. (Cambridge, England, 1890), I, 377f. Maxwell regarded the statistical method as less satisfactory than "the dynamical method" which strictly determined the behavior of individual particles by

means of the Lagrangean and Hamiltonian equations of motion. Modern physics has departed from Maxwell's view and followed the line of Peirce's and Boltzmann's statistical conception of law.

5. See J. Herschel, "Quetelet on Probabilities," *Edinburgh Review,* XLII, 1–57 (1850); Quetelet's work of collecting and methodizing the statistics of crime and suicides in different countries was utilized by Buckle and Mill. H. T. Buckle's *History of Civilization in England,* 3 vols. (Oxford, 1857–1904), quotes Quetelet in vol. I, p. 22: "In every thing which concerns crime, the same numbers re-occur with a constancy which cannot be mistaken; and this is the case even with those crimes which seem quite independent of human foresight, such, for instance, as murders, which are generally committed after quarrels arising from circumstances apparently casual. Nevertheless, we know from experience that every year there not only takes place nearly the same number of murders, but that even the instruments by which they are committed are employed in the same proportion." J. M. Keynes, in his *A Treatise on Probability* (London, 1921), pp. 334–335, regards Quetelet's influence to be due more to his sensational language than to an accurate logic of probability.

6. Chauncey Wright, "Sir Charles Lyell (1797–1875)," *Nation,* XX, 146–147 (March 4, 1875).

7. *Life and Letters of Thomas H. Huxley* (New York, 1901), I, 168.

8. Chauncey Wright, "Who Are Our Ancestors?" *Nation,* XX, 405–407 (June 17, 1875). One of the possible sources of Wright's idea of the neutrality of science with respect to religious sentiment was Faraday's Sandemanian attitude; see Sylvan Thompson, *Michael Faraday, His Life and Work* (New York, 1898). Henry James, Sr., had adopted Faraday's Glasite faith after meeting Faraday in England through their mutual friend, Joseph Henry. Dr. Oliver Wendell Holmes observed that Francis Galton's *Hereditary Genius* (1869) had shown that "there is a frequent correlation between an unusually devout disposition and a weak constitution," E. M. Tilton, *Amiable Autocrat, A Biography of Dr. Oliver Wendell Holmes* (New York, 1947), p. 324.

9. Edward Hitchcock, *The Religion of Geology and Its Connected Sciences* (1st ed.; London, 1851, Boston, 1856), p. 433.

10. Hitchcock, p. 439. Hitchcock defended the catastrophic theory despite the criticisms of Charles Lyell's *Principles of Geology* (1830–1833).

11. Ernst Haeckel, *Natural History of Creation* (5th ed., translated and revised by E. Ray Lankester, 1911), II, 450.

12. Haeckel, II, 458.

13. Haeckel, II, 451.

14. Hitchcock, p. 454.

15. Hitchcock, p. 463.

16. Hitchcock, p. 472.

17. Hitchcock, p. 507.

18. Hitchcock's authorities are Dr. Chalmers, *Works,* VII, 262; Sir John Herschel, *Discourse on the Study of Natural Philosophy;* Sir John P.

Smith, *Lectures* (4th ed.), p. viii; McCosh, *Method of Divine Government,* pp. 449f.; and Sedgwick, *Discourse on the Studies of the Universities* (5th ed.), p. 105 (appendix). Hitchcock regarded "the stony volume of the earth's history" as a means man would probably possess in the future of reading the past history of the world and of individuals' acts of wickedness or virtue (p. 438). His pious geology also invoked the idea of the millennium: "We know, also, from the joint testimony of Scripture and geology, that another change is to pass over the world, to prepare it for inhabitants far more elevated than those now living upon it, and in possession of perfect holiness and perfect happiness. And it may be it will experience far greater changes, adapting it for higher and higher grades of being, through periods of duration to which we can assign no limits. O, what a vast chain of being is here spread out before the imagination, reaching immeasurably far into the depths of the eternity which is past, and into the eternity which is to come!"

19. W. M. Smallwood, *Natural History and the American Mind* (New York, 1941), p. 238.

20. Smallwood, chapter viii: The Philosophy of the Naturalist. See also, H. T. Pledge, *Science Since 1500* (London, 1939), p. 158: "Men's minds were full of the great new laws of physics; but natural selection turned out to be not a law but a litany—sung over the graves of those who were not fit. Neither Darwin nor anyone else at the time saw fully that 'fitness,' taken in a broad enough sense to cover all the cases, merely *means* survival. Probably these men's ideas of fitness came from the human case."

21. *Calendar of the Letters of Charles Darwin to Asa Gray,* Historical Records Survey (Boston, 1939), letters 35, 56, 97, 103, 132, 133.

22. Paul Ansel Chadbourne, *Instinct: Its Office in the Animal Kingdom and Its Relation to the Higher Powers in Man,* Lowell Lectures, 1871 (2d ed.; New York, 1883), p. 20. A footnote on page 155 critically but respectfully refers to Wright's view of "accidental variations" in his critical review of Mivart's *Genesis of Species* (*North American Review* of July 1871).

23. Chadbourne, p. 14.

24. Chadbourne, pp. 23–25. Cf. Wright's extensive correspondence with Darwin on "unconscious selection" as a factor in the evolution of dialects. *Letters of Chauncey Wright, with some account of his Life,* edited by James Bradley Thayer (Cambridge, Mass., 1878), pp. 240f.; this volume is hereafter referred to as *C. W.* Darwin's important letter of June 3, 1872, quoted in part in a footnote (*ibid.,* p. 240), is among the Northampton papers of Chauncey Wright. When Darwin asked Wright: "As your mind is so clear and as you consider so carefully the meaning of words, I wish you would take some incidental occasion to consider when a thing may properly be said to be effected by the will of man," he had been prompted to the wish by Professor Whitney's philological quarrel with Schleicher on the development of language by conscious rules. John Fiske's earliest article was on the evolution of language. The whole purport of pragma-

tism, according to Peirce and James, was to clarify the language of meta-physics and establish operational criteria of meaning. The lawyers of the Metaphysical Club, Green and Holmes, devoted much attention to legal semantics. Cf. C. S. Peirce on the "ethics of terminology," *C.P.,* 2.219–226, 5.413, 5.610.

25. Chauncey Wright, "The Evolution or Self-Consciousness," *North American Review* (April 1873) reprinted in Charles Eliot Norton, *Philosophical Discussions by Chauncey Wright with a Biographical Sketch of the Author* (New York, 1877), p. 235.

26. Chadbourne, p. 59.

27. Chadbourne, p. 70.

28. Chadbourne, p. 72.

29. Chadbourne, p. 73.

30. L. Agassiz, *Methods of Study in Natural History* (Boston, 1863), p. 281.

31. Chadbourne, pp. 75–76.

32. Chadbourne, p. 77.

33. Chadbourne, p. 47.

34. Chadbourne, p. 83.

35. Chadbourne, pp. 294–95.

II. THE BIRTHPLACE OF PRAGMATISM: PEIRCE'S METAPHYSICAL CLUB

1. William James, *Pragmatism: A New Name for Some Old Ways of Thinking* (New York, 1907). The book was dedicated "To the Memory of John Stuart Mill from Whom I First Learned the Pragmatic Openness of Mind and Whom My Fancy Likes to Picture As Our Leader Were He Alive To-day." In the chapter on "What Pragmatism Means" (pp. 43ff.), James refers not only to Peirce's essay on "How To Make Our Ideas Clear" (*Popular Science Monthly,* 1878) but also to the chemist Ostwald's "Theorie and Praxis" (*Zeitschrift des Oesterreichischen Ingenieur und Architecten-Vereines,* 1905, nrs. 4, 6). James quotes a letter to him from Ostwald who, in his lectures on the philosophy of science, was using the principle of pragmatism without the name: "All realities influence our practice and that influence is their meaning for us. I am accustomed to put questions to my classes in this way: In what respects would the world be different if this alternative or that were true? If I can find nothing that would become different, then the alternative has no sense . . . [Many a] controversy would never have begun if the combatants had asked themselves what particular experimental fact could have been made different by one or the other view being correct. For it would then have appeared that no difference of fact could possibly ensue." James goes on to mention as old hands at the pragmatic method: Socrates, Aristotle, Locke, Berkeley, Hume; among his contemporaries, he mentions Shadworth Hodgson, Sigwart, Mach, Pearson, Milhaud, Poincaré, Duhem, Heymans, Papini, Schiller, and Dewey.

2. William James, "Philosophical Conceptions and Practical Results,"

University of California Chronicle (1898), a lecture delivered before the Philosophical Union at Berkeley, California, in 1897. Reprinted in *Collected Essays and Reviews*, edited by R. B. Perry (New York, 1920), p. 410.

3. The only two steady positions Peirce ever had as regular sources of income were as an instructor in logic at Johns Hopkins University (1879–1884) and as aid, computer, and assistant, successively, for the United States Coast and Geodetic Survey (1859–1891). His poverty at Milford, Pennsylvania, where he spent the last third of his hermit-like life (1891–1916) is described in Paul Weiss's article on Peirce in the *Dictionary of American Biography*, XIV (1934), 398–403. There is further evidence in a letter of his brother, James M. Peirce, to President Daniel Gilman of Johns Hopkins University asking that Charles Peirce be invited to give lectures on the history of science. Sporadic lecturing, book reviewing, hack-work for dictionaries, and translating scientific articles from French and German for the Smithsonian Institution were his only sources of income after 1891. See "The Peirce-Langley Correspondence etc. (At the Smithsonian Institution)," edited by P. P. Wiener, *Proceedings of the American Philosophical Society*, XCI, 201f. (1947), especially the letter of Peirce to Secretary Langley, April 20, 1901, p. 208: "If your Disbursing Agent would send that $4 it would pay my fare to N. Y. and return. I have had to pay off more than half my mortgage this spring amounting to $1700 and had to dump every cent I got from you into that. Then I had a suit which would bring me $800. Unfortunately one of our 'Lay Judges' had a suit against the same party, and thought that if I got all that, there would not be enough left for him, and he managed iniquitously so that I only get $200, not enough to pay expenses hardly. I can appeal; but meantime, I am just as hard up as can be, with Mrs. Peirce's operation coming shortly." Several persons who knew Peirce in the 1900's have testified to me orally of his destitution then.

4. Ralph Barton Perry, *The Thought and Character of William James*, 2 vols. (Boston: Little, Brown, and Co., 1935), provides the definitive life and intellectual biography of William James (1842–1910), and of his pragmatism. Professor Perry believes that the Club "in which James was associated with Peirce and Chauncey Wright was most active in the years 1870–1872" (II, 407). Wright was in Europe from July to November 1872, during which time he paid a visit to Darwin.

5. Catherine Drinker Bowen, in *Yankee from Olympus* (Boston, 1944), the latest life of Justice Holmes (1841–1935), refers to the Metaphysical Club (p. 253) which "flourished in the early '70's" (p. 427), but does not give the source (probably Perry). The best account of Holmes's pragmatism is Max Fisch's "Justice Holmes, the Prediction Theory of Law, and Pragmatism," *Journal of Philosophy*, XXXIX, 85–97 (1942). But Dr. Fisch errs in making John Chipman Gray (1839–1915), lifelong friend of James and Holmes, a member of the Metaphysical Club instead of the more formal dining club to which James and Holmes belonged at the time (cf. Perry, I, 360). He also errs in setting the date 1869 for the formation of the Metaphysical Club, on insufficient evidence. However, I agree with

Dr. Fisch's important observation about the Club (p. 88): "The one thing that all seven [?] had in common, beside a Harvard degree, was an enthusiasm for the British tradition in philosophy, and a sense of the epoch-making importance of Darwin's *Origin of Species*."

6. Joseph B. Warner, Harvard A.B. 1869, LL.B. 1873, died in 1923 without having published any work, but had assisted Holmes in editing Kent's *Commentaries on American Law*. See Appendix G.

7. Like Warner, Green was a Harvard Law School graduate (1851). See Appendix F. The editor of Chauncey Wright's letters, James B. Thayer, notes that "in later years, that acute but learned lawyer, the late Nicholas St. John Green (afterwards a Professor in the Boston Law School) whom he [Chauncey Wright] had first known as a student at Northampton, and whose permanent home was in Cambridge, came gradually to be intimate with Chauncey" (*C. W.*, pp. 31–32).

8. Alexander Bain, *The Emotions and the Will* (3d. ed.; London, 1875), p. 505. Chauncey Wright used Bain's text in his course on psychology at Harvard in 1870; see *C. W.*, p. 202, and for Wright's high opinion of Bain's analysis of æsthetic pleasures and of their "extensive intellectual affinities," see *ibid.*, p. 297 and pp. 178–179.

9. *Query:* Is it Bain or Green who is regarded by Peirce as "the grandfather of pragmatism"? R. B. Perry (II, 407) says Bain, but P. Weiss (*DAB*, article, "Peirce, C. S.") says Green.

10. Since Chauncey Wright (born 1830 in Northampton, Massachusetts) died in 1875, Peirce's statement implies that the Club never met after September 12, 1875. Neither of the biographers, who were also close friends, of Wright, mentions the Metaphysical Club (cf. *C. W.*, and Wright, *Philosophical Discussions*).

11. See Sir Leslie Stephen, *The English Utilitarians*, vol. III: John Stuart Mill (London, 1900). Stephen was also the author of *Essays on Freethinking and Plain Speaking* (1873) and *The Agnostic's Apology and other Essays* (London, 1893), a defense of Humean skepticism and positivism. Thomas H. Huxley created the term "agnosticism" in 1869. According to Stephen, "crude empiricism was transformed into evolutionism" by the union of utilitarianism and Darwin's theories (Mill, III, 375). Pragmatism, we may say, tried to consummate this union in diverse fields.

12. See Chapter III.

13. John Fiske (1842–1901) delivered a series of lectures on evolution a few years before they were published as *Outlines of Cosmic Philosophy, based on the Doctrine of Evolution, with Criticisms on the Positive Philosophy*, 2 vols. (London, 1874). This book was discussed one evening in 1874 by Wright, Peirce, St. John Green, Warner, and James, and Fiske, reports James to T. Sargent Perry (*Letters of William James*, edited by Henry James, Jr., 2 vols. (Boston, 1920), II, 233). Van Wyck Brooks confuses this group with one of the other clubs to which James and Holmes belonged (*New England: Indian Summer*, p. 256). See Chapter VI.

14. F. E. Abbot (1863–1903), classmate of Charles Peirce at Har-

vard, began his metaphysical debut with a bold criticism of Kant and Hamilton in two articles, "The Philosophy of Space and Time" and "The Conditioned and the Unconditioned," in the *North American Review* (July and October 1864), which formed the first topics of a lengthy correspondence with Chauncey Wright lasting from 1864 to 1869 (cf. *C. W.*, pp. 56ff). After leaving his post as Unitarian minister at Dover, New Hampshire, in 1867 because of his "radical" views on evolution, Abbot wrote an article which Wright had declined to do for Abbot's "Free Religious Association," namely "Positivism in Theology" (*Christian Examiner* [March 1866]). Abbot vigorously attacked Spencer's mechanistic *Principles of Biology* in a long review entitled "Philosophical Biology" (*North American Review* [October 1868], pp. 377–422). The vitalistic organicism of Abbot's interpretation of Darwinian evolution was elaborated in his *Organic Scientific Philosophy: Scientific Theism* (Boston, 1885), including a defense of a "scientific realism" based on the "objectivity of relations." This book, highly praised by Peirce, was occasioned by a lecture Abbot gave before the Concord School of Philosophy, July 30, 1885, in a symposium on the question: "Is Pantheism the Legitimate Outcome of Modern Science?" The other symposiasts were John Fiske, W. T. Harris, A. P. Peabody, G. H. Howison, and Edward Montgomery. Abbot's largest philosophical treatise is *The Syllogistic Philosophy or Prolegomena to Science*, 2 vols. (begun in 1893 and finished in 1903). Cf. Appendix G, notes 11–16.

15. *C. P.*, 5.12 (*c.* 1906). The six volumes of Peirce's papers do not follow chronological order, but Justus Buchler's *Charles Peirce's Empiricism* (New York, 1939), appendix ii, supplies dates. The index item "Autobiographical" of each volume of *C. P.* and Weiss's biographical article in the *DAB*, XIV, 398–403, are also invaluable to the future historian of Peirce's thought. Peirce says that he deliberately refrained as late as 1893 from including the term "pragmatism" among the many philosophical items and terms he wrote for the *Century Dictionary*, because he did not believe the term had been sufficiently defined and accepted as philosophical currency (*C. P.* 5.13). The published papers do not give an adequate idea of the extent of Peirce's contributions to the history of thought. His unpublished MSS. include twelve lectures, delivered in 1892, on the "History of Science" (from the Babylonians to Newton); plans for an edited translation of Petrus Peregrinus *On The Lodestone;* the concept of natural law in eighteenth-century England at the time Hume wrote the "Essay on Miracles"; comparative biography and experiments on judgments of greatness of men; "The Productiveness of the Nineteenth Century in Great Men"; a history of logic; British Logicians of the Nineteenth Century. Weiss does not discuss these in his *DAB* article, and also flatly asserts: "Pragmatism, Peirce's creation, had its origin in the discussions, in Cambridge, of a fortnightly [*sic*] 'metaphysical club' founded in the seventies."

16. In 1865 when Mill published his *Examination of the Philosophy of Sir William Hamilton.* See Wright's review of this book in *Nation*, I, 278f. (1865), "Mill on Hamilton." Also see an unpublished letter of C. S. Peirce

to Wright, dated September 2, 1865, among the papers of Chauncey Wright
in the possession of Mrs. D. P. Abbot of Northampton, Mass. In this letter,
Peirce writes:

"I have read Mill's book. His Free Will discussion is highly ingenious.
His chapter on Mansell is the most striking and though I distrust such emphatic philosophizing I think in this case Mill is right. I would like to have
Mansell reply; as he must do or give up the game. The contradictions in
Hamilton are well brought out; but with a malicious intent. Mill wants to
root out this philosophy, by adequate arguments or by inadequate ones.
Hamilton's thoughts for 20 years are spread out in these posthumous publications and what man's thoughts for so many years is self-consistent? Only
a dolt's. Mill does not seem to me a greater thinker than Hamilton. He has
been most careful not to publish anything that he might afterwards wish to
contradict; and when he finds Sir W. H.'s position so exposed as it is, to the
great benefit of the world and by his editors rather than by himself—Mill
mercilessly assails it. These *ad hominem* arguments are not contributions to
philosophy, but they will have a great effect on the public. In his criticism on the Logic, Mill said just about what I expected.

"I have invented a little trick at cards . . ." [Then follow five pages
of two mathematical card-tricks, with Pyramus and Thisbe as dramatis
personae.]

17. *Journal of Philosophy,* XIII, 719–30 (1916).

18. *Peirce MSS.* at Widener Library, I B I, box 1: "Pragmatism Made
Easy."

19. "Pragmatism Made Easy," p. 2.

20. "Grounds of the Validity of the Laws of Logic," "Questions Concerning Certain Faculties Claimed for Man," and "Some Consequences of
Four Incapacities." Reprinted in *C. P.,* vol. V: Pragmatism and Pragmaticism.

21. I am indebted to Dr. Joseph Blau for a photostat copy of this fragment given to him by Mr. Philip P. Chase of Milton, Massachusetts.

22. "The Holmes-Cohen Correspondence," *Journal of the History of
Ideas,* IX, 19 (1948).

23. "Holmes-Cohen Correspondence," p. 35; letter of Holmes to
Cohen, September 14, 1923: "That we could not assert necessity of the
order of the universe I learned to believe from Chauncey Wright long ago.
I suspect C. S. P. [Peirce] got it from the same source."

24. *C. P.,* 6.482. "A Neglected Argument for the Reality of God,"
Hibbert Journal, VII, 90–112 (1908).

25. *C. P.,* 5.414. "What Pragmatism Is," *Monist,* XV, 161–181
(1905).

26. Letter to Mrs. Ladd-Franklin; *Journal of Philosophy,* XIII, 718
(1916).

27. Perry, II, 409.

28. Perry, II, 407n.5.

29. *Kritik der reinen Vernunft* (Zweiter Auflage) in *Immanuel Kant's*

Sämtliche Werke in Sechs Bänden (Leipzig, 1922), Dritter Band, pp. 620f. (Methoden lehre, II Hauptst., III Absch.), my translation. Kant distinguished three kinds of rules in his *Foundation of the Metaphysic of Morals:* rules of fitness ("regeln der Geschicklichkeit . . . *technisch* zur Kunst gehörig"), counsels of prudence ("Ratschläge der Klugheit . . . *pragmatisch* zur Wohlfahrt"), and commands of morality (*"Gebote . . . Gesetze der Sittlichkeit"*). Only the last, the categorical imperative, is purely ethical. "Denn pragmatisch werden die *Sanktionen* genannt, welche eigentlich nicht aus dem Rechte der Staaten als notwendige Gesetze, sondern aus der *Vorsorge* für die allgemeine Wohlfahrt fliessen. Pragmatisch ist eine *Geschichte* abgefasst, wenn sie *klug* macht, d.i. die Welt Belehrt, wie sie ihren Vorteil besser, oder wenigstens eben so gut als die Vorwelt besorgen könne" (*Kant's Sämtliche Werke,* V, 44).

30. *C. P.,* 5.412.

31. Holmes to Pollock, August 30, 1929; *Holmes-Pollock Letters: the correspondence of Mr. Justice Holmes and Sir Frederick Pollock, 1874–1932,* edited by Mark DeWolfe Howe, 2 vols. (Cambridge, Mass., 1941), II, 252.

32. *Letters of John Fiske,* edited by Ethel Fiske Fisk (New York: The Macmillan Company, 1940).

33. Brooks, *New England: Indian Summer,* pp. 111n., 276.

34. Arthur O. Lovejoy, "The Thirteen Pragmatisms," *Journal of Philosophy,* V, 1–12, 29–39 (1908).

35. *C. W.,* p. 124; letter to F. E. Abbot, October 28, 1867.

III. CHAUNCEY WRIGHT, DEFENDER OF DARWIN AND PRECURSOR OF PRAGMATISM

1. Asa Gray, "Darwin and His Reviewers," *Atlantic Monthly,* VI, 423 (1860). It is altogether likely that Wright's own key conception of the neutrality of science owed much to Gray and to Jeffries Wyman; cf. *C. W.,* pp. 42, 271, 286, 331, 367.

2. Darwin's letter to Wright (July 14, 1871) is among the Darwin-Wright letters at Northampton.

3. *North American Review,* vol. CXII (1873). Wright's article is reprinted in *Philosophical Discussions,* pp. 199–266; cf. p. 259: "It becomes an interesting question, therefore, when in general anything can be properly said to be effected by the will of man." Wright's analysis makes the will of man an effective agency only when it operates under the influence of the same natural conditions as we find in the geologic effects of winds, rains, and rivers or in the biological effects of natural selection; but in matters of moral and legal actions, the ethics of responsibility has evolved from the primitive tribal leader's authority to new powers of tradition, language, and other arbitrary customs. This sort of analysis is quite different from the admittedly "crass-supernaturalism" of James's individualistic will to believe.

4. Chauncey Wright, "On the Architecture of Bees," *Proceedings of the American Academy of Arts and Sciences,* IV, 432 (May 8, 1860). Bowen's four letters to Wright are at Northampton. On Bowen's controversy with Dr. Holmes, see Tilton, *Amiable Autocrat,* pp. 264, 424 n. 11.

5. *C. W.,* p. 43; letter to Mrs. Lesley, February 12, 1860. Agassiz' *Essay on Classification* had appeared in 1857.

6. *C. P.,* 5.64.

7. *C. W.,* pp. 218–19; letter to Mrs. Lesley, May 7, 1871. For Darwin's acknowledgment of Wright, see *Descent of Man* (New York, 1871) II, 319.

8. *C. W.,* pp. 162–63; letter to Miss Grace Norton, January 13, 1870.

9. McCosh's Scottish common-sense intuitionism attempted to reconcile Calvinism and Darwin by admitting spontaneous variations and natural adaptation as direct results of supernatural intervention by acts of special providence. See H. W. Schneider, *A History of American Philosophy* (New York, 1946), pp. 370f.; Chauncey Wright, "McCosh on Intuitions," *Nation* (1865), p. 627.

10. Chauncey Wright, "German Darwinism," *Nation,* XXI, 168 (Sept. 9, 1875).

11. "Nihilism" was a term used by Turgenev in *Fathers and Sons,* but was introduced into philosophy as a protest against "the ontological passion" by David Masson in his *Recent British Philosophy* (New York, 1866), reviewed by Wright in the *Nation,* III, 385–86 (Nov. 15, 1866). Cf. Henry James's review of Turgenev's pessimistic novels in the *North American Review,* vol. CXVIII (1874).

12. See Perry, vol. I, ch. xix: Depression and Recovery.

13. Perry, II, 718–721: appendix iii, Notes by Chauncey Wright.

14. *C. W.,* pp. 328–330.

15. C. S. Peirce, "Hume on Miracles and Laws of Nature," in the Peirce-Langley Correspondence and Manuscripts, edited by P. P. Wiener, *Proceedings of the American Philosophical Society,* XCI, 219 (1947). See Appendix D, pp. 223–226, above, for a summary of Peirce's analysis of Hume's ideas. Langley did not publish Peirce's essay, twice revised, because of a mutual misunderstanding.

16. *C. W.,* pp. 274f.; letter to Miss Grace Norton, July 29, 1874.

17. William James, "Chauncey Wright," *Nation,* XXI, 194 (Sept. 23, 1875). Reprinted in James's *Collected Essays and Reviews.* Cf. *Letters of William James;* letter of March 1869 to Thomas W. Ward: "I'm swamped in an empirical philosophy." The editor notes: "About this time a man whose thought was vigorously materialistic was often at the house in Quincy Street. This was Chauncey Wright." But Wright condemned materialism as an unenlightening metaphysic, and a matter of temperament. Cf. Appendix A.

18. Perry, I, 525–528.

19. Perry, II, 720.

20. Perry, II, 718, quoted from James's obituary notice on Chauncey Wright, *Nation,* XXI, 194 (Sept. 23, 1875).

21. *C. W.,* pp. 97f.; letter to Charles Eliot Norton, February 18, 1867.

22. *C. W.*, p. 100; letter to F. E. Abbot, July 9, 1867.

23. Letter of C. S. Peirce to William James, October 3, 1904, in Perry, II, 431–32.

24. Letter to William James, dated Paris, November 21, 1875, in Perry, I, 537. Peirce's suggestion in this letter that he would like "to give some résumé of Wright's ideas and of the history of his thought" was not adopted by either C. E. Norton, who edited Wright's *Philosophical Discussions*, or J. B. Thayer who edited Wright's letters.

25. See *Hale* v. *Everett*, 53 *New Hampshire Reports* 1.

26. See Sidney Warren, *American Free Thought, 1860–1914* (New York, London, 1943).

27. *C. W.*, p. 107; letter to C. E. Norton, July 24, 1867: "I have had the pleasure of meeting Mr. Abbot, my metaphysical antagonist, personally, and debating in a pleasant conversation the various points of our controversy. I found him as able in talk as in his writings." Cf. letter to Miss Jane Norton, March 22, 1869, *C. W.*, pp. 144f.

28. *C. W.*, p. 56; letter to F. E. Abbot, December 20, 1864.

29. *C. P.* 5.223, fn. 2. Peirce had also defended Hamilton against Mill's *Examination* in the letter to Chauncey Wright, September 2, 1865, quoted in note 16, of Chapter II, above.

30. *C. W.*, p. 61.

31. *C. W.*, p. 111; letter to F. E. Abbot, August 13, 1867.

32. *C. W.*, pp. 111–12.

33. *C. W.*, pp. 112–13.

34. *C. W.*, p. 125; letter to F. E. Abbot, October 28, 1867.

35. *C. W.*, p. 126.

36. *Nation*, XIII, 355 (1871).

37. *C. W.*, p. 132.

38. *C. W.*, p. 133.

39. *C. W.*, p. 134.

40. *C. W.*, p. 140; letter to F. E. Abbot, February 10, 1869.

41. *C. W.*, p. 141.

42. *Ibid.*

43. *C. W.*, pp. 141–42.

44. Abbot, *Organic Scientific Philosophy: Scientific Theism*, p. 218.

45. H. W. Schneider, "The Influence of Darwin and Spencer on American Philosophical Theology," *JHI*, VI, 18 (Jan. 1945).

46. *North American Review*, CXIII, 63–103 (July 1871). Mivart's book was one of four books reviewed: A. R. Wallace's *Contributions to the Theory of Natural Selection* (1870), Mivart's *On the Genesis of Species* (1871), Darwin's *Descent of Man*, 2 vols. (1871) and *On the Origin of Species*, 5th ed. (1871). Wright devoted most of his review to criticizing Mivart with polite respect but unrelenting insistence on Mivart's failure to understand the neutrality of scientific method in general and of Darwin's hypothesis of natural selection in particular, based on long accumulated specific evidence, rather than on speculation about cosmic evolution. Only the latter could give rise to moral and religious qualms.

47. *Quarterly Review* (London), CXXXI, 89–90 (July 1871).

48. "Evolution by Natural Selection," *North American Review*, CXV, 3–4 (July 1872).

49. Quoted by Bert J. Loewenberg, "The Reaction of American Scientists to Darwinism," *American Historical Review*, XXXVIII, 691 (1933).

50. "Evolution by Natural Selection," p. 3.

51. Cf. John Fiske, *Darwinism and Other Essays* (Boston, 1879), "Darwinism Verified," pp. 1–14. In the second chapter, "Mr. Mivart on Darwinism" (Fiske's review of Mivart's *Lessons from Nature*, New York, 1876), Fiske came to the defense of his friend Wright: "Mr. Mivart, in reprinting his rejoinder to Mr. Chauncey Wright, takes care not to inform the reader of the surrejoinder which came from his powerful antagonist" (p. 33). Fiske attributed Wright's failure to receive any public notice to the fact that he was "singularly devoid of literary ambition which leads one to seek to influence the public by written exposition. Indeed, had he possessed more of this kind of ambition, perhaps the requisite knack would not have been wanting; for Mr. Wright was not deficient in clearness of thought or command of language. The difficulty—or, if we prefer so to call it, the esoteric character—of his writings was due rather to the sheer extent of their richness and originality. His essays and reviews were pregnant with valuable suggestions . . . An intellect more powerful for its union of acuteness with sobriety has not yet been seen in our country. In these respects, he reminds one of Mr. Mill whom he so warmly admired"; *Harvard Advocate*, reprinted in Fiske's *Darwinism and other Essays*, pp. 75–104. Cf. Henry James's tribute to Wright's character and philosophical promise, Appendix A.

52. "Evolution of Self-Consciousness," *North American Review*, CXVI, 245–310 (April 1873).

53. "Evolution of Self-Consciousness," p. 249.

54. This letter is included in the Northampton papers of Chauncey Wright. Darwin wrote to Alfred Russel Wallace about Wright's review of Mivart: "The article though not very clearly written, and poor in parts from want of knowledge, seems to me admirable"; *Life and Letters of Charles Darwin*, edited by his son (New York, 1904), II, 324.

55. "On the Use and Arrangement of Leaves," *Memoirs of the Academy of Arts and Sciences* (Oct. 10, 1871). In a postscript to the letter of July 14, 1871, Darwin adds: "I have been struck by your allusion (p. 98) to Phyllotaxy. I have spent days in puzzling over it, and quite unhappy from not being a mathematician. I supposed new leaves to be added to or removed from (i.e., to stand closer or farther from each other) an existing plant, and then that the leaves had to be arranged anew, so as to be equally exposed to light. I assumed that from the spiral vascicles being in bundles, it was necessary after a time that one leaf should stand almost over another. —I did not take much regard to packing in the bud, as Nägeli says that young leaves are sometimes irregularly placed in the bud, and subsequently place themselves at regular angles. Now with your view explain why there

are angles (as I found by drawing them) by which the leaves can be arranged at equal distances from each other with one leaf standing over another at some distance above and below, *and yet that such angles never occur in nature?* If you can explain this, I wish you would be induced to publish on the subject.—C. D." In another postscript, it is interesting to note, Darwin informs Wright: "I am beginning to think that plants are more wonderful than animals!"

56. *Runkle's Mathematical Monthly,* I, 244–248 (1859): "The Most Thorough Uniform Distribution of Points About an Axis." See also a letter from J. E. Oliver to Wright (May 1, 1859), at Northampton.

57. Three volumes appeared (1859, 1860, 1861) up to the Civil War. The editor, Runkle, was Wright's employer in the United States Naval Observatory at Cambridge, where for three months a year Wright worked as a computer in order to earn his annual salary fast enough to give the rest of the year to his more cherished philosophical studies.

58. *Goethes Naturwissenschaftliche Schriften* in *Goethe's Werke,* II. Theil, 7. Band: Zur Morphologie (Weimar, 1892), pp. 37–68.

59. Julius von Sachs, *History of Botany (1530–1860),* translated by H. E. F. Garnsey, revised by I. B. Balfour (Oxford, 1890). Also see Link's "Report on Physiological Botany for 1844–5" in *Reports and Papers on Botany of the Ray Society* (London, 1849). Sachs criticizes the platonizing idealism of the German advocates of Phyllotaxis as "a construction which was geometrically correct but which could hardly be made to agree with the history of development and the mechanical forces concerned."

60. Sachs, p. 169. Note how the faithful disciple and biographer of Louis Agassiz refers to Chauncey Wright: "A few words on Chauncey Wright, and his singular similarity to another adversary of Agassiz, Karl Schimper [an antivitalist], will not be out of place. Chauncey Wright was a mathematician of talent, who turned his mathematical skill [in defense of Darwin] to a study of the phyllotaxis of plants, just as Schimper had done forty years before. Agassiz treated Wright in the most friendly way, even appointing him a lecturer [in philosophy] at his school for girls, just as he had treated Karl Schimper. Wright was an earnest seeker for truth, but he was above all a great dreamer, and some of his writings are rather obscure"; Jules Marcou, *Life, Letters, and Works of Louis Agassiz* (New York, 1896), II, 31.

61. M. J. Schleiden, *The Plant: A Biography* (London, 1853), p. 91.

62. Cf. *Charles Darwin, Autobiography and Letters,* edited by his son, Francis Darwin (New York, 1893), pp. 292–93.

63. Perry, I, 520ff.

64. *North American Review,* vol. CXI (Oct. 1870): "Limits of Natural Selection," a review of Alfred Russel Wallace's *Contributions to the Theory of Natural Selection.* Cf. *C. W.,* p. 271.

65. *Nation,* vol. XX (April 22, 1875): "McCosh on Tyndall."

66. *Nation,* XIII, 59–60 (July 27, 1871). A further example of Wright's philosophical and speculative temporalism may be cited from one

of many similar utterances: "Time is always an element in phenomena, both internal and external, and is the abstract of relations in general. It is the *continuum* of phenomenal existences. It is their 'substance' in the logical sense, that is, their universal attribute"; Perry, II, 720–21.

67. *Philosophical Discussions*, p. 407. Cf. William James's view that metaphysics is a matter of temperament. In this review of Mill's *Examination* of the philosophy of Hamilton, Wright says that Mill is separated from Hamilton's Kantianism "by that fundamental division . . . throughout the whole history of speculative thought; a division which is at bottom one of feeling and mental character, and one in which no love is lost"; *Nation*, I, 279 (Aug. 31, 1865).

68. *Philosophical Discussions*, preface, p. xvii. Also, for Wright's attitude to metaphysics, cf. *C. W.*, pp. 270f.

69. *Nation*, vol. XX (April 22, 1875): "McCosh on Tyndall."

70. *Philosophical Discussions*, p. 324. Wright adduced the Ptolemaic notion of "saving the appearances" and Newton's "hypotheses non fingo" as further historical illustrations of the metaphysical neutrality of science. Cf. Wright's "The Genesis of Species," *North American Review*, vol. CXIII (July 1871), which impressed Charles Darwin so much that he had it reprinted and circulated in England at his own expense.

71. "German Darwinism," *Nation*, XXI, 169 (Sept. 9, 1875).

72. "Mill on Comte," *Nation*, vol. II (Jan. 4, 1866).

73. "McCosh on Intuitions," *Nation*, I, 627–629 (Nov. 16, 1865).

74. Letter of Wright to Darwin, February 24, 1875; *C. W.*, p. 335.

75. "The word Darwinism has become as familiar as Galvanism or Mormonism"; review of "Darwin's *Variation of Animals and Plants under Domestication*," in the *Nation*, VI, 234–236 (1868).

76. *Ibid.*

77. Review of A. R. Wallace's *Contribution to the Theory of Natural Selection*, in the *North American Review* (Oct. 1870).

78. *Nation*, II, 724–25 (June 8, 1866).

79. *Nation*, XX, 113–14 (Jan. 1875), and *Nation*, XXI, 168–170 (Sept. 9, 1875). Asa Gray referred to this review in one of his letters as "an article which was to have been continued by Chauncey Wright, in which he points out clearly the essential difference between Darwinism, which is scientific, and Spencerism, which is 'philosophical.' Save the mark! —Poor Wright died suddenly of apoplexy, Sunday morning. He was a staunch Millite, and very acute and clear-headed" (*Letters of Asa Gray*, edited by Jane Loring Gray, 2 vols. (Boston, 1893), II, 657: letter of September 14, 1875). Gray, who wrote the first systematic textbook on botany in the United States for use in colleges (1st ed., 1842), had modified his Linnaeism and theism in favor of Darwin. It is noteworthy that the opening sentence of the first edition which quotes "the immortal Linneus" on minerals, vegetables, and animals, was dropped from the later editions of his *Botanical Text-Book*, and that phyllotaxis is not discussed until the

second edition (1845), pp. 123–127, and at greater length in the third edition (1850), pp. 140–150.

80. *Nation*, XXI, 169 (Sept. 9, 1875). When W. T. Harris submitted a paper on "Herbert Spencer" to the *North American Review*, Norton asked Wright for an opinion. Wright's reply that it was "the mere dry husk of Hegelianism,—dogmatic without the only merit of dogmatism, distinctness of definition" (*C. W.*, p. 87), provoked Harris to issue *The Journal of Speculative Philosophy* as the organ of the St. Louis school of Hegelians, with his paper on Spencer as the starting article of the first number (Jan. 1877). There is no doubt that Wright, had he lived longer, would have regarded Peirce's speculations on evolutionary love as a species of German Darwinism. For Peirce regarded as hopeless Wright's lifelong effort to synthesize Mill's associationalism (in which matter was the "possibility of sensations") with the "living facts of Darwinian development"—hopeless, in Peirce's metaphysical mind, which rises from the immediacy of the Firstness of sensation to higher categories in Schellingian and Hegelian fashion. This sort of philosophizing was thoroughly repugnant to Wright's more modest empiricism.

81. Cf. J. B. Stallo, *The Concepts and Theories of Modern Physics* (New York, 1st ed. 1881, 2d ed. 1884), ch. xv: "Cosmological and Cosmogenetic Speculations.—The Nebular Hypothesis" (2d ed.), pp. 270–293.

82. *Philosophical Discussions*, p. 5.

83. *Philosophical Discussions*, p. 7.

84. *Ibid.* Cf. B. Russell, who in *Scientific Method in Philosophy* (Oxford, 1914), ch. 1, applies the same sort of criticism to evolutionary theology as Wright did; for example, in his review of Spencer's *Biology* (1866), *Nation*, II, 724 (June 8, 1866).

85. *Philosophical Discussions*, p. 7.

86. *Philosophical Discussions*, p. 9.

87. See note 34, Ch. II above.

88. Chauncey Wright, "Mill on Hamilton," *Nation*, I, 278f. (Aug. 31, 1865).

89. *C. P.*, 6.291f.: "The *Origin of Species* merely extends politico-economic views of progress to the entire realm of animal and vegetable life" (from Peirce's "Evolutionary Love," *Monist*, III, 176f. [1893]). Curiously enough, Marx expressed a similar view of Darwin's work in a letter to Engels; however, Marx here accepted the argument from analogy, whereas Wright and Peirce rejected it.

90. Simon Newcomb, *Reminiscences of an Astronomer* (Boston and New York, 1903), p. 408. Cf. *C. W.*, pp. 71f., in which Wright agrees with Newcomb on the same determinateness of effects holding for the voluntary actions of men as for all other natural changes.

91. *C. W.*, p. 263; letter to E. L. Godkin, June 3, 1874.

92. *C. W.*, pp. 148–152.

93. *C. W.*, p. 174; letter to C. E. Norton, March 21, 1870.

94. *C. W.,* p. 174.
95. Charles Sumner Letters, Widener Library, Harvard University.
96. *C. W.,* pp. 38, 47, 50, 51.
97. *C. W.,* p. 183.
98. *C. W.,* p. 184.
99. *C. W.,* p. 181.
100. *C. W.,* p. 179.

IV. THE EVOLUTIONISM AND PRAGMATICISM OF PEIRCE

1. *New Republic* (Feb. 3, 1937), p. 416. John Dewey had previously written an account of Peirce's and James's versions of pragmatism in an article for the special American number of the *Revue de métaphysique et de morale,* XXIX, 411–430 (Oct. 1922), translated from the French in the supplement, "Development of American Pragmatism," *Studies in the History of Ideas,* II, 351–377 (1925). Dewey there properly indicates the European background of Peirce's thought in relation to Kant's critical idealism. The present chapter is intended to provide more data from the unpublished manuscripts of Peirce, which were not accessible to Dewey, concerning the evolution of Peirce's peculiar variety of pragmatism. Cf. Charles Hartshorne, "Charles Sanders Peirce's Metaphysics of Evolution," *New England Quarterly,* XIV, 49–63 (1941).

2. See Chapter II above.

3. Unpublished Peirce manuscripts in the Harvard University Archives, Widener Library, catalogued by C. I. Lewis and K. McMahon.

4. Peirce's interest in Semiotic is as early and persistent in his intellectual history as his interest in scientific logic. He planned a philosophical thesaurus to supplant Roget's, and contributed definitions and analyses of philosophical terms to the *Century Dictionary* and Baldwin's *Dictionary of Philosophy.*

5. Though Peirce's dissatisfaction with the relations among Kant's categories is shared by many post-Kantians who attempted to unify them, the important difference is that Peirce's critique went back to the defects of Aristotle's logic and its forms of judgment, and emerged with a scientifically fruitful logic of relations, whereas the post-Kantians (Schelling, Hegel, *et al.*) produced a metaphysical and transcendental dialectic far beyond scientific logic. Yet Peirce toward the end of the century considered his philosophy "Schellingism transformed in the light of modern physics" (letter to William James, Jan. 28, 1894, in Perry, II, 415–16).

6. *C. P.,* 2.654.

7. "The Unconscious," especially in Von Hartmann's *Philosophy of the Unconscious (Philosophie des Unbewussten,* 1869), looms large in German metaphysics and psychology after Schopenhauer. Cf. William James, *Principles of Psychology,* 2 vols. (New York, 1890), I, 162f.

8. Perry, I, 231. Peirce says of his studies in 1867–68: "My work became self-controlled in the year 1867, paper of May 14: 'On a New List

of Categories'—grown and developed in the intervening years. It was not until the next year I think I took up seriously the Logic of relations"; letter to Cousin Jo, June 1909 ("Folder of Late Fragments," *Peirce MSS.*, Box IB 3a, p. 15).

9. Perry, I, 229. Peirce's nine "Lectures on British Logicians" (1869) are among the unpublished Peirce manuscripts at Widener, I B2, Box 10. Cf. Perry, I, 321. In 1868, the *Journal of Speculative Philosophy*, founded by W. T. Harris of the St. Louis school of post-Kantian idealists, published four important articles of Peirce's in which he attacked intuitionism and upheld a social ideal of truth. See note 20, Chapter II above.

10. *North American Review*, vol. CXIII (1871). When Peirce placed Berkeley among the ancestors of pragmatism, it was not because of the nominalistic sensationalism or psychological analyses in the Bishop's theory of knowledge, but rather because of the acute logical and methodological contributions of the latter's critique of the foundations of Newtonian mechanics. It should be remembered that Berkeley's *Analyst*, published in 1734, played a role in the clarifying of the fundamental ideas of the infinitesimal calculus prior to Robin's *Discourse Concerning the Nature and Certainty of Sir Isaac Newton's Method of Fluxions and of Prime and Ultimate Ratios* (1735), Maclaurin's *Treatise of Fluxions* (1742), and Carnot's *Réflexions sur la métaphysique du calcul infinitésimal* (1797), and that Mach in his historical and critical works on mechanics paid high tribute to Berkeley. As samples of the kind of basic logical and methodological questions raised by Berkeley, the following may be cited:

"I have no controversy about your conclusions, but only about your logic and method: how do you demonstrate? what objects you are conversant with, and whether you conceive them clearly? What principles you proceed upon; how sound they may be; and how you apply them? . . . And what are these fluxions? The velocities of evanescent increments. And what are these same evanescent increments? They are neither finite quantities, not quantities infinitely small, nor yet nothing. May we not call them the ghosts of departed quantities?

"You may possibly hope to evade the force of all that hath been said, and to screen false principles and inconsistent reasonings, by a general pretence that these objections and remarks are *metaphysical*. But this is a vain pretence . . . Qu. 4. Whether men may properly be said to proceed in a scientific method, without clearly conceiving the object they are conversant about, the end proposed, and the method by which it is pursued? . . . Qu. 8. Whether the notions of absolute time, absolute place, and absolute motion be not most abstractly metaphysical? Whether it be possible for us to measure, compute, or know them?" These passages are quoted from Berkeley, *The Analyst: A Discourse Addressed to an Infidel Mathematician*, in David Eugene Smith, *A Source Book in Mathematics* (New York, 1929), pp. 633–34. F. Cajori who edited these selections notes that the "infidel mathematician" to whom Berkeley addressed his discourse was supposed to be Edmund Halley, the celebrated astronomer (1656–1742), though there is no evidence of religious skepticism in Halley's published writings (D. E. Smith, p. 627).

11. *Nation,* XIII, 355 (1871). Cf. Wright's review of Fraser's edition of Berkeley, *Nation,* XIII, 59 (1871).

12. *Nation,* XIII, 386 (1871).

13. "Hume on Miracles and Laws of Nature" (revised by Peirce, after correspondence with Langley, to "The Laws of Nature and Hume's Argument Against Miracles") an unpublished manuscript by Peirce (1901) in the files of the Smithsonian Institution at Washington, D. C. See Appendix D.

14. "German Darwinism," *Nation,* vol. XXI (Sept. 9, 1875).

15. *C. P.,* 1.22: "Aristotle . . . whose system, like all the greatest systems, was evolutionary." Also cf. 1.173: "Aristotle's philosophy . . . is but metaphysical evolutionism" (*c.* 1897).

16. *C. P.,* 1.104 (*c.* 1896).

17. *C. P.,* 1.105 (*c.* 1896).

18. *C. P.,* 1.33 (1869).

19. *C. P.,* 5.366 ("The Fixation of Belief," 1877). Peirce added a footnote to this passage in 1903: "Let us not, however, be cocksure that natural selection is the only factor of evolution; and until this momentous proposition has been much better proved than as yet it has been, let it not blind us to the force [of] very sound reasoning."

20. *C. P.,* 6.297; cf. also 6.293.4 (1893).

21. *C. P.,* 1.203ff. (1902).

22. *C. P.,* 1.205 (fn. 1); cf. also 1.229–231 (1902).

23. Agassiz, *Methods of Study.* These Lowell Lectures stated more popularly the ideas of the *Essay on Classification,* but were directed against Darwin.

24. *C. P.,* 1.571 (1910).

25. *Ibid.*

26. *C. P.,* 1.5 (*c.* 1897). Peirce refers to what he calls the "pseudo-evolutionism" of Spencer, because it is based on mechanical law, in 6.157 (1892). Cf. also 6.14 (1891).

27. *C. P.,* 6.604 (1893). Cf. *C. W.,* pp. 364n., 366.

28. *C. P.,* 1.103 ("Lessons from the History of Science," the last of twelve lectures delivered in 1892–93, all written and among the Peirce manuscripts at Widener; they form a book planned by Peirce on The History of Science, from the Babylonians to Newton).

29. *C. P.,* 6.14 (1891). That Peirce's "thorough-going evolutionism" implied for him a philosophical way of getting into "personal relations with God," appears in 6.157 (1892).

30. *C. P.,* 6.15 (1891). See the critical comment on this passage in Professor Lovejoy's "Notes on Peirce's Evolutionism," Appendix E.

31. *Peirce MSS.,* IB 3a "Folder of Late Fragments" ("Why should the Doctrine of Chances raise Science to a Higher Plane?" (Jan. 24, 1909), p. 15).

32. *Ibid.,* p. 16.

33. *Johns Hopkins University Circulars* (May 1881), I, 150.

34. *C. P.,* 1.396 (*c.* 1890).

35. *C. P.,* 5.364 ("The Fixation of Belief," *Popular Science Monthly,* XII, 1–15 [Jan. 1877]). John Venn's *Logic of Chance* appeared in 1866; F. Galton, a Lamarckian, inferred the stability of heredity from the existence of "independent causes" fulfilling the normal law of distribution of variations (*Philosophical Magazine,* XLIX, 44 [1875]); Herschel, who influenced Mill, regarded the normal mean as our only ground for confidence that the future would resemble the past (*Edinburgh Review,* XCII, 23 [1850]). Venn spoke of the statistical "life-span" of man as "Nature's aim, the length of life for which she builds a man" (*Journal of the Statistical Society* (1891), p. 443). Cf. also K. Pearson's "Contributions to the Mathematical Theory of Evolution," *Transactions of the Royal Society* (1894–1903), and Peirce's scathing review, "Pearson's Grammar of Science," *Popular Science Monthly,* LVIII, 296–306 (1900).

36. Cf. *The Life of James Clerk Maxwell, with Selections from his Correspondence and Occasional Writings,* edited by L. Campbell and W. Garnet (London, 1884), p. 362. Maxwell regarded "dynamical laws" as more genuine forms of explanation than statistical laws in which the individual is not determined. See Chapter I, note 4, above.

37. *C. P.,* 6.430ff. ("A Religion of Science," *Open Court,* VII, 3559–60 [1893]).

38. *C. P.,* 6.33 (1891).

39. *C. P.,* 1.7 (c. 1897).

40. *C. P.,* 5.423 ("What is Pragmatism?" *Monist,* XV, 161–181 [1905]).

41. *Ibid.*

42. *Peirce MSS.,* "On Positivism," *c.* 1860.

43. *Ibid.* On Abbot's scientific theism, see Schneider, "Influence of Darwin and Spencer," *JHI,* VI, 15–18 (Jan. 1945).

44. Letter of C. S. Peirce to Cousin Jo (June 1909), p. 15 in Box IB 3a, "Folder of Late Fragments," *Peirce MSS.*

45. Prior to his Johns Hopkins' *Studies in Logic* (1883), Peirce had as a physicist in the employ of the United States Coast and Geodetic Survey written ten reports, published as appendices to that bureau's annual reports. The first was "On the theory of errors of observation" (Report of 1870, appendix 21), and the third was "Theory of the economy of research" (Report of 1876, appendix 14). The latter anticipates Mach's ideas on the economy of scientific thought.

46. *C. P.,* 2.86 (1902). Cf. William James, "Great Men, Great Thoughts and the Environment," *Atlantic Monthly,* XLVI, 441f. (1880), for a view similar to that of Peirce concerning evolution in scientific thinking by inductive elimination of hypotheses. See Chapter VI, section 1.

47. *C. P.,* 3.526, 527 (1897).

48. *Philosophical Discussions,* p. 219.

49. *Philosophical Discussions,* p. 221. Note the Lamarckian idea of inheriting acquired habits, an idea widely shared during the latter part of the nineteenth century by biologists (Haeckel, Romanes, Cope, Gray, and even Darwin at times) as well as by philosophers and literary people (Wright, Peirce, Samuel Butler, Ibsen). Gray, who was the teacher of both Wright and Peirce, declared, when he came out for Darwin in a long review of "Darwin on the Origin of Species": "The assertion that acquired habitudes or instincts, and acquired structures, are not heritable, any breeder or good observer can refute," *Atlantic Monthly,* VI, 423 (1860).

50. *C. P.,* 4.9 (1906).

51. *C. P.,* 3.608 (*c.* 1903).

52. *C. P.,* 3.620 (1902).

53. Quoted from Peirce (*C. P.,* 5.253) by John Dewey, "Peirce's Theory of Linguistic Signs, Thought, and Meaning," *Journal of Philosophy,* XLIII, 88 (Feb. 14, 1946).

54. "Peirce's Theory of Linguistic Signs," p. 92; *C. P.,* 2.220.

55. "Peirce's Theory of Linguistic Signs," p. 91; *C. P.,* 2.293.

56. George Boole (1815–1864), *The Mathematical Analysis of Logic, being an Essay towards a Calculus of Deductive Reasoning* (1847); *An Investigation of the Laws of Thought, on which are founded the Mathematical Theories of Logic and Probability* (1854). Attention should be called here to the subtitle of the latter, Boole's principal work, and to the purpose of that work as expressed on the first page: "The design of the following treatise is to investigate the fundamental laws of those operations of the human mind by which reasoning is performed; to give expression to them in the symbolic language of a Calculus, and upon this foundation to establish a science of Logic and construct its method; to make that method itself the basis of a general method for the application of the mathematical doctrine of Probabilities; and finally, to collect from the various elements of truth brought to view in the course of these inquiries some probable intimations concerning the nature and constitution of the human mind."

57. Unpublished Peirce manuscript at the Smithsonian Institution, pp. 14–15 of autoscript copy. See Appendix D.

58. *C. P.,* 5.4 (1902).

59. *Ibid.*

60. *C. P.,* 5.402 ("How To Make Our Ideas Clear," 1878).

61. *C. P.,* 5.395 ("How To Make Our Ideas Clear").

62. *C. P.,* 5.407 ("How To Make Our Ideas Clear").

63. In an introduction attached in 1903 to a combined version of "The Fixation of Belief" and "How To Make Our Ideas Clear"—entitled "My Plea for Pragmatism"—Peirce judged Royce's Gifford Lectures (1900–01) on *The World and the Individual* "a work not free from faults of logic, yet valid in the main" (*C. P.,* 5.358n.). Peirce's references to Henry James, Sr., occur in *C. P.,* 5.402 n.3, 6.287, 6.507, citing *Substance and Shadow* (1863) very favorably.

64. *C. P.,* 6.157 (1892).

65. *C. P.,* 1.348 (1903). Note the shift from the contingency of our knowledge of laws to the evolution of the laws themselves.

66. *C. P.,* 1.487 (*c.* 1896).

67. Lovejoy, *Great Chain of Being,* pp. 324–25.

68. *C. P.,* 5.433 (1905); also cf. 1.615 (1903).

69. *Peirce MSS.,* "Folder of Late Fragments," I B 3a.

70. *Ibid.,* letter to Cousin Jo (June 1909).

71. See Appendix E, "Note on Peirce's Evolutionism" by Professor Lovejoy, part of his epistolary communication to the present writer, on the logical validity of Peirce's tychistic argument.

V. DARWINISM IN JAMES'S PSYCHOLOGY AND PRAGMATISM

1. Perry, I, 469.

2. Cf. Ronald B. Levinson, "A Note on One of James's Favorite Metaphors," *JHI*, VIII, 237–239 (1947). The metaphor of the *salto mortale*, Mr. Levinson indicates, occurs in the following marginal note written by James about 1875 in his copy of Sigwart's *Logik* (I, 368): "This postulate of a real world amounts to nothing more than the execution of a *salto mortale* from subjectivity to objectivity and the 'will to believe' that our ideas transcend themselves and reveal a reality whose form of self-consistency is the same as that of our logical thinking." Santayana, a pupil of James, used the idea in his *Scepticism and Animal Faith: Introduction to a System of Philosophy* (New York, 1923).

3. James, *Collected Essays and Reviews,* pp. 411–12, quoted by R. B. Perry, *In the Spirit of William James* (New Haven, 1938), p. 59.

4. Perry, *Thought and Character,* I, 476.

5. James, *Principles of Psychology,* preface, p. vi.

6. *Educational Review,* I, 371 (1891).

7. *International Journal of Ethics,* I, 371 (1891).

8. *International Journal of Ethics,* I, 148 (1891).

9. *Mind,* XVI, 396 (1891).

10. Letter of William James to James Ward, dated Florence, November 1, 1892, in Perry, *Thought and Character,* II, 96–97.

11. James, *Principles of Psychology,* vol. I, ch. ii.

12. James, *Psychology,* I, 79. Professor Lovejoy, in a letter to me, has suggested that this doctrine of James "is akin to that later set forth by Bergson, that (a) consciousness is primary; that (b) it somehow *produces* brain and nervous system as means to its own ends; that (c) for its purposes, it is better that some of the cerebral or neural centres should become automatic—or non-automatic—i.e., should go on operating without constant control by conscious desires or purposes—but that others should become determined, not merely by desires, but by intelligence." For a further, critical comment on Bergson's evolutionism, see R. G. Collingwood, *The Idea of Nature* (Oxford, 1945), pp. 136ff. On James and Bergson see Georges Sorel, *De l'Utilité du Pragmatisme* (Paris, 1917), pp. 2ff.

13. James, *Psychology*, I, 80.

14. James, *Psychology*, I, 82.

15. James, *Psychology*, I, 109.

16. James, *Psychology*, I, 110, 117–120, 157, 374, 419; II, 522.

17. Another contemporary source of James's ideo-motor theory is Dr. Isaac Ray's *Treatise on the Medical Jurisprudence of Insanity* (5th ed.; Boston, 1871) which we shall come across, in Chapter VII below, as a work reviewed by Nicholas St. John Green, the brilliant lawyer friend of James, Wright, Peirce, Fiske, and Holmes. See also Gregory Zilboorg, *A History of Medical Psychology* (New York, 1940), p. 415, on the need of scientific psychology in legal cases involving plea of insanity. Zilboorg, on pages 375ff. discusses Janet's and Charcot's theories of the "unconscious" automatism of the pathological mind, theories which influenced James's own conception of "secondary selves."

18. Cf. Chapter I, section 3 above, and James, *Psychology*, II, 383f.

19. James, *Psychology*, I, 384.

20. *Ibid.*

21. James, *Psychology*, II, 386.

22. James, *Psychology*, II, 391n.

23. James, *Psychology*, II, 392.

24. James, *Psychology*, II, 401.

25. James, *Psychology*, II, 441.

26. James, *Psychology*, II, 409–10.

27. James, *Psychology*, II, 428.

28. James, *Psychology*, II, 430–31.

29. James, *Psychology*, II, 438.

30. James, *Psychology*, II, 410.

31. James, *Psychology*, II, 429.

32. James, *Psychology*, II, 426.

33. James came out as early as 1880 in defense of Darwin against Lamarck in his essay, "Great Men, Great Thoughts and the Environment," *Atlantic Monthly*, XLVI, 444 (Oct. 1880).

34. Cf. E. B. Titchener, "An Historical Note on the James-Lange Theory of Emotion," *American Journal of Psychology*, XXV, 427–447 (1914). On Lotze and James, see O. Kraushaar, "Lotze's Influence on the Pragmatism and Practical Philosophy of William James," *JHI*, I, 439–458 (1940).

35. James, *Psychology*, II, 447.

36. James, *Psychology*, II, 449.

37. James, *Psychology*, II, 452.

38. James, *Psychology*, II, 478.

39. James, *Psychology*, II, 479.

40. James, *Psychology*, II, 480n.

41. James, *Psychology*, II, 484.

42. In 1905 after reading G. Heyman's *Einführing in der Metaphysik*, William James told his metaphysical friend and critic, Charles A. Strong,

that he had become a convert to panpsychism; but the conversion was a short lived one. Cf. Perry, *Thought and Character*, II, 403f.

43. James, *Psychology* (ch. vi: The Mind-Stuff Theory), I, 145.

44. James, *Psychology*, I, 161f.

45. James, *Psychology*, I, 182.

46. James, *Psychology*, II, 494.

47. James, *Psychology*, II, 565. James's footnote refers to Aristotle's *Nicomachean Ethics*, vii, 3, for the discussion of incontinence, and to Sir A. Grant's edition (2d ed.), I, 212, for the discussion of the doctrine of "The Practical Syllogism."

48. *C. W.*, p. 342.

49. *C. W.*, p. 343.

50. *C. W.*, p. 330. Chauncey Wright's sociological use of the adverb "pragmatically" in the passage is the earliest occurrence of the term in the literature, so far as I know. It is noteworthy that the context is a critical account by Wright of the historical use of the Platonic theory of "natural rights." In Charlotte Brontë's *The Professor* (London, 1856), the word "pragmatical" occurs to characterize a meddlesome reformer's opinions.

51. James, *Psychology*, II, 651.

52. James, *Psychology*, II, 335fn. in which James quotes from his own article, "The Sentiment of Rationality," *Mind*, vol. IV (1879).

53. James, *Psychology*, II, 667.

54. James, *Psychology*, II, 668.

55. James, *Psychology*, II, 669.

56. James, *Psychology*, II, 671. Cf. Peirce's idea of "the growth of concrete reasonableness."

57. James, *Psychology*, II, 670fn. in which James refers to Chauncey Wright's critique of the scholastic attack on Darwin in Mivart's *Genesis of Species*.

58. James, *Psychology*, II, 688–89 (last paragraph of the book).

59. Address by William James, "Report of the Fifth Annual Meeting of the New England Anti-Imperialist League" (Boston, Nov. 28–30, 1903), pp. 21–26.

60. *Ibid.*, p. 26.

61. Cf. Perry, *Thought and Character*, II, 289f.; Merle Curti, *Social Ideas of American Educators* (New York, 1935), 438.

62. *Letters of William James*, II, 100.

63. "The Will to Believe" was originally delivered before the philosophical clubs of Yale and Brown Universities, and published in the *New World*, June 1896. Reprinted in *The Will to Believe, and Other Essays in Popular Philosophy* (London, 1897), pp. 1–31. James dedicated the book "To My Old Friend, Charles Sanders Peirce, To whose philosophic comradeship in old times and to whose writings in more recent years I owe more incitement and help than I can express or repay."

64. *Ibid.*, penultimate paragraph.

65. *C. P.*, 6.184.

VI. SOCIAL EVOLUTION AND FISKE'S PHILOSOPHY OF HISTORY

1. *Atlantic Monthly,* XLVI, 441 (Oct. 1880). The essay is entitled "Great Men and Their Environment" in *The Will to Believe,* pp. 216–254.

2. *Atlantic Monthly,* XLVI, 442 (Oct. 1880).

3. *Ibid.* See a parallel passage in Nicholas St. John Green's criticisms of an unbroken mechanical chain of causation in nature, "a pure fabrication of the mind," in "Proximate and Remote Cause," *American Law Review* (1870), quoted (in our next chapter) from Green's *Essays and Notes on the Law of Torts and Crime,* edited by Frederick Green (Menasha, Wis., 1933), pp. 14–15.

4. *Atlantic Monthly,* XLVI, 444 (Oct. 1880).

5. *Atlantic Monthly,* XLVI, 445 fn. 1 (Oct. 1880), in which James refers to Chauncey Wright's *Philosophical Discussions* (p. 165) "for some striking remarks on the different orders of magnitude with which the different phenomenal kinds of force act." The idea is that a minute variation may in time and through the operation of natural selection produce great changes.

6. *Atlantic Monthly,* XLVI, 445 (Oct. 1880). In *The Will to Believe,* pp. 225–26.

7. See Matthiessen, *The James Family,* p. 460, ·for William James's expressions of admiration for Carlyle's individualism in contrast with Henry James Senior's antagonism to Carlyle's lack of "spiritual socialism." See also Schneider's chapter on Henry James, Sr., "Spiritual Socialism and Spontaneity," in *History of American Philosophy,* pp. 301ff.

8. *Atlantic Monthly,* XLVI, 457 (Oct. 1880).

9. John Fiske, "Sociology and Hero-Worship: An Evolutionist's Reply to Dr. James," *Atlantic Monthly,* XLVII, 82 (Jan. 1881). Grant Allen, the British Spencerian, defended his social Lamarckism against James in "The Genesis of Species," *Atlantic Monthly,* XLVII, 371–381 (March 1881).

10. *Atlantic Monthly,* XLVII, 84 (Jan. 1881).

11. Fiske, *Outlines of Cosmic Philosophy,* II, 188.

12. *Cosmic Philosophy,* II, 163. Fiske was greatly influenced by Buckle's *History of Civilization in England,* vol. I (London, 1857), ch. iii, which condemns the metaphysician's method as "the direct opposite of the historical method; the metaphysician studying one mind [introspectively], the historian studying many minds."

13. *Cosmic Philosophy,* II, 166.

14. *Cosmic Philosophy,* II, 169. (Fiske refers here to Froude's *Short Studies on Great Subjects,* vol. I.)

15. *Cosmic Philosophy,* II, 174. Berkeley's complaint against metaphysical dust-raising is also cited by Buckle (*Civilization in England,* fn. 135 in ch. iii of vol. I).

16. *Cosmic Philosophy,* II, 179 n. 1, refers to Bain's *The Emotions*

and the Will, (1st ed.; London, 1859), p. 550. This work of the positivist Bain was used as the text of Chauncey Wright's university lectures on psychology in 1870, and was also well known to Green, Peirce, James, and Holmes. Holmes met Bain and Mill in London in 1866, and while impressed with their knowledge of fact, he noted in his diary that they lacked the infinite perspective of metaphysics.

17. See W. Stull Holt, "The Idea of Scientific History in America," *JHI,* I, 352–362 (1940). "Both because of its position in nineteenth-century thought and because of its obvious connection with mankind, biology was the science to which the historians most frequently turned. It furnished them with a terminology which they used again and again in their historical writing. In his famous essay on 'The Significance of the Frontier in American History,' which may well have been suggested by an application of Darwinism to history, Turner quickly stated his case in biological terms" (pp. 352–53). "Those of us," wrote Henry Adams, "who read Buckle's first volume when it appeared in 1857, and almost immediately afterwards, in 1859, read the *Origin of Species* and felt the violent impulse which Darwin gave to the study of natural laws, never doubted that historians would follow until they had exhausted every possible hypothesis to create a science of history." Quoted by Holt (p. 354) from Henry Adams, "The Tendency of History," *Annual Report of the American Historical Association for 1894,* pp. 17–18. Holt shows that another group of American historians followed the positivistic theory of the German historian von Ranke (p. 355), and concludes that "the previously current conceptions of 'scientific history' must be modified" (p. 362). The critical study of the preconceptions of the positivism and evolutionism of the latter part of the nineteenth century would certainly support Professor Holt's conclusion.

18. *The Letters of John Fiske,* edited by his daughter Ethel F. Fisk (New York, 1940), p. 41; letter of June 24, 1860. The following letter of Fiske to Spencer (Feb. 20, 1864) narrates the transition from Comtist to Spencerian evolutionary positivism: "At the time [1861] when I reviewed Buckle I was just passing out from Comtism. During six months of incessant study and reflection my former idols were all demolished. Having successively adopted and rejected the system of almost every philosopher from Descartes to Prof. Ferrier, I began the year 1860 with Comte, Mill, and Lewes. I then favored the scheme of acquiring a general knowledge of all the sciences in their hierarchical order as laid down by Comte, which scheme was eventually carried out. I first noticed your name in Mr. Lewes's little exposition of Comte early in 1860, and the extract from *Social Statics* there given led me to put down my name for *First Principles* . . . The influence of your writings is apparent alike in every line of my writings and every sentence of my conversation . . . I graduated at Harvard last summer and am now connected with the University as a student of Law. It is my purpose to occupy the leisure time left by my profession in working out a complete theory of the origin and evolution of language after the manner sketched in my Essay on that subject.

"Associated with me to some extent in my studies, and endeavoring to carry out the same principles into Jurisprudence, is Mr. George Roberts, an attorney in the office of Mr. Justice Carter." *The Life and Letters of John Fiske*, edited by John Spencer Clark, 2 vols. (Boston and New York, 1917), I, 293–94. Spencer in his reply to Fiske repudiated the current idea that he was a Comtist. *Ibid.*, I, 296.

19. *Letters*, p. 73; letter of November 10, 1861.

20. *Letters*, p. 106; letter of September 19, 1863. Benjamin Peirce told J. J. Sylvester that Fiske, "one of our greatest philosophical thinkers . . . without any special study of astronomy, has done more for the nebular hypothesis than either you or I could have done." Clark, *Life and Letters*, II, 68. The mathematician Benjamin Peirce, himself not a very good expositor of science for the untrained public, is obviously alluding to Fiske's success as a popular writer and lecturer on evolution. Fiske and his friend Youmans, founder and editor of *Popular Science Monthly*, were not merely promoters of Spencer's philosophy, but they also brought science into adult education. Cf. Charles M. Haar, "Edward Livingston Youmans: a Chapter in the Diffusion of Science in America," *JHI*, IX, 193–213 (April 1948).

21. J. Fiske, "The Genesis of Language," *North American Review* (Oct. 1869), pp. 324ff.

22. John Fiske, *A Century of Science, and Other Essays* (New York, 1899), "The Scope and Purport of Evolution" [March 1890], p. 40.

23. *Letters*, p. 185; letter of July 5, 1869, in which Fiske takes credit for having helped Eliot's election to the presidency of Harvard in 1868. On Fiske's difficulties with the Overseers prior to 1868, see F. Greenslet, *The Lowells and Their Seven Worlds* (Boston, 1946), p. 312.

24. *Letters*, p. 118; letter of January 3, 1864.

25. *Letters*, p. 336; letter of Darwin to Fiske, December 8, 1874.

26. Clark, *Life and Letters*, I, 458f. With Darwin, Fiske discussed Wright's ideas on phyllotaxis or the arrangements of leaves around the stems of plants. *Ibid.*, I, 460.

27. Clark, *Life and Letters*, I, 460.

28. John Fiske, *The Unseen World* (Boston, 1876), p. 18. This book went into twelve editions.

29. *Unseen World*, p. 53.

30. *Unseen World*, p. 52.

31. *Unseen World*, p. 143.

32. *Unseen World*, pp. 145–46.

33. *Cosmic Philosophy*, I, 272. The italics are mine in the passage quoted.

34. *Ibid.*, n. 1. In referring to G. H. Lewes' *Aristotle*, p. 92, and *Problems of Life and Mind*, I, 317, is Fiske as well as Lewes mistaking Osiander's preface to the *Revolutionibus* as Copernicus'? The Thorn edition of Copernicus' *De Revolutionibus* appeared in 1873. It was not until 1858 that Kepler's exposure of Copernicus' disagreement with Osiander's preface was

printed; see E. Rosen, "The Ramus-Rheticus Correspondence," *JHI*, I, 366, n. 22 (1940).

35. *Cosmic Philosophy*, I, 272.

36. *Cosmic Philosophy*, I, 273.

37. *Cosmic Philosophy*, I, 275.

38. *Cosmic Philosophy*, II, 252.

39. *Cosmic Philosophy*, II, 267–68.

40. *Cosmic Philosophy*, II, 232. Fiske refers to Comte's *La Philosophie positive*, vol. V, for "a marvelous tableau of the progress of society" (II, 237). At the end of his work, Fiske credits Comte with a more historical approach to society than De Maistre's.

41. *Cosmic Philosophy*, II, 233.

42. *Cosmic Philosophy*, II, 226.

43. *Cosmic Philosophy*, II, 227 (in which Fiske quotes Spencer's *Essays*, 2d series, p. 154).

44. *Cosmic Philosophy*, II, 227–28.

45. *Cosmic Philosophy*, II, 228.

46. Richard Hofstadter, *Social Darwinism 1860–1915* (Philadelphia, 1944), p. 151. Cf. J. Fiske, "Who Are the Aryans?" *Atlantic Monthly* (Feb. 1881), pp. 224–234: "It is never safe to use language as a direct criterion of race, for speech and blood depend on different sets of circumstances, which do not always vary together . . . During the past twenty five years [1855–1880] Frenchmen [Taine, Gobineau?] have had a good deal to say about the 'Latin race.' There could hardly be a more flagrant instance of the perversion of a linguistic name to ethnological purposes" (*ibid.*, pp. 231–32).

47. J. Fiske, *The Destiny of Man Viewed in the Light of his Origin* (1884), p. 89.

48. *Cosmic Philosophy*, II, 290.

49. Cf. S. J. Holmes, *Life and Morals* (New York, 1948), p. 110, on the prolongation of infancy in anthropoid apes as one of "the deep roots of altruism . . . the evolutionary importance of which has been emphasized by John Fiske."

50. J. Fiske, *Discovery of America, with some Account of Ancient America, and the Spanish Conquest* (1892), I, 59. Other references in Fiske's works to his theory of prolonged infancy occur in *Outlines of Cosmic Philosophy*, pt. II, chaps. xvi, xxi, xxii; *Excursions of an Evolutionist*, pp. 306–319; *Darwinism and other Essays*, pp. 40–49; *Destiny of Man*, sections iii-ix.

51. *Discovery of America*, I, 59 fn. 2.

52. *Discovery of America*, I, 23.

53. *Discovery of America*, I, 32.

54. *Discovery of America*, preface vi-vii.

55. Thomas Sargent Perry, *John Fiske* (Boston, 1906), p. 86.

56. Chapter vi: The Critical Attitude of Philosophy, *Cosmic Philosophy*.

57. *Cosmic Philosophy,* II, 483.
58. *Cosmic Philosophy,* II, 479.
59. *Cosmic Philosophy,* II, 485.
60. *Cosmic Philosophy,* II, 475.
61. *Cosmic Philosophy,* II, 483–84.
62. *Cosmic Philosophy,* II, 500.
63. *Cosmic Philosophy,* II, 501.

VII. THE PRAGMATIC LEGAL PHILOSOPHY OF
NICHOLAS ST. JOHN GREEN

1. O. W. Holmes, Jr., "The Profession of the Law," lecture delivered to the undergraduates of Harvard on February 17, 1886, in *Speeches* (Boston, 1913), p. 23, and in *Collected Legal Papers,* edited by H. J. Laski (New York, 1920), p. 30.

2. *C. P.,* 5.12.

3. John Chipman Gray in his *Nature and Sources of the Law* (New York, 1909) stresses the positivistic view that the law is whatever the courts, at different places or times, lay down for the determination of legal rights and duties. On Gray's relation to Holmes and James, see Fisch, "Justice Holmes, the Prediction Theory of the Law, and Pragmatism," *Journal of Philosophy,* XXXIX, 85–97 (1942), and my note in *JHI,* VII, 219 (1946).

4. *C. W.,* p. 367. See also Wright's note on Sir Henry Maine's *Ancient Law and History of Early Institutions* in the *Nation,* XXI, 9 (July 1, 1875). Only a year after graduation, Wright showed an interest in legal thought at Harvard: "In the Law School there is a vigor of thought and a stimulus to study which can't be found elsewhere" (*C. W.,* p. 33; letter to George H. Fisher, Dec. 4, 1853).

5. *American Law Review* (Oct. 1876), quoted in the editorial preface to *Essays and Notes on the Law of Tort and Crime* by Nicholas St. John Green, pp. v–vi. Cf. review in *Harvard Law Review,* XLVII, 555.

6. Nicholas St. John Green, *Criminal Law Reports,* 2 vols. (Boston, 1874–75), vol. I, preface.

7. *American Law Review,* XV, 332 (May 1881).

8. *American Law Review,* IV, 350–353 (March 1870). This review is not included in Green's *Essays and Notes.* It is mistakenly credited to O. W. Holmes, Jr., by Silas Bent, *Justice Oliver Wendell Holmes, a Biography* (New York, 1931). It is listed as Green's in Jones's *Index to Legal Periodical Literature,* vol. I (Boston, 1888).

9. *American Law Review,* VIII, 649 (1874), reprinted in Green's *Essays and Notes,* pp. 93f.

10. *Essays and Notes,* pp. 1–17.

11. *Essays and Notes,* p. 1.

12. *De Augmentis,* 82d Aphorism.

13. The texts cited by Green are: Aristotle's *Post. Analytics* (bk. I, ch. 13), Zabarella's *Opera* (416b. 835f.), *Artis Logicae Rudimenta* (Al-

drich, Mansel's ed. p. 118), Ficino's *Theologia Platonica* (bk. IX, ch. 4), Joannes Versor's *Quaestiones super novam logicam* (i. an. post.), *Averroes' Commentary on the Post. Analytics* (lib. I, ch. 2, comment. 30), Petrus Tartaretus' *Expositio super text. logices Aristotelis,* Contus' *In Universam Dialecticam Aristotelis* (1 anal. post., cap 10), Ockham's *Summa Totius Logicae* (2d pt. of 3d pt. ch. 20), Duns Scotus' *Questiones Subtillissimae in Metaphysicam* (bk. V, question 1; *Opera* IV, pp. 595b. 596b.), Aquinas' *Summa Theologica* (1st pt. of 2d pt. ques. 79, art. 1), Burgersdyk's *Institutiones logicae* (bk. I, ch. 17, ax. ch. 15, 18, 34, 35), Eustachius a Sancto Paolo's *Summa Philosophiae Quadripartita* (pt. 3, p. 39). Of Duns Scotus, Green says (pp. 9–10) that he "achieved a fame bounded only by the limits of the civilized world. He died in his thirty-fourth year, the intellectual giant of the time. His works in bulk are equal to those of some of the law writers of the present day, although he does not appear to have written anything for the sole purpose of swelling their size. They are extant in twelve folio volumes. His clearness, depth, and power of mind would put to the blush the bold ignorance of those who speak patronizingly of scholastic subtlety, if happening to possess the capacity to understand logical statement and close reasoning, they should perchance read a page of his writing. He was the greatest of the British schoolmen." Peirce and Abbot also spoke highly of Duns Scotus' subtlety as a logician and metaphysician, but Green wished only to show Bacon's scholasticism and its *irrelevance* to modern law.

14. *Essays and Notes*, p. 11.
15. *Essays and Notes*, p. 14.
16. *Essays and Notes*, pp. 15–17.
17. *Essays and Notes*, p. 11.
18. *Essays and Notes*, pp. 14–15.
19. *Essays and Notes*, p. 163.
20. *Essays and Notes*, pp. 16–17.
21. *American Law Review*, V, 704 (1871), reprinted in *Essays and Notes*, pp. 161–169.
22. *Essays and Notes*, p. 166.
23. *Essays and Notes*, p. 165.
24. *Essays and Notes*, p. 53.
25. *Ibid.*
26. *Ibid.*
27. *Essays and Notes*, pp. 66–67.
28. Twelfth edition, I,[21], fn. 1 (b).
29. *Essays and Notes*, p. 56.
30. *American Law Review*, XIV, 1–35 (Jan. 1880).
31. *American Law Review*, XIV, 9, 23 (Jan. 1880).
32. *Essays and Notes*, pp. 165, 192.
33. *Common Law*, pp. 62, 110.
34. *Essays and Notes*, p. 67. See notes 27 and 28, above, the latter being direct evidence of Green's influence on Holmes.

35. *American Law Review,* XIV, 21 (Jan. 1880).
36. *Essays and Notes,* p. 161.
37. Green, *Essays and Notes,* p. 163; Holmes, *American Law Review,* XIV, 22 (Jan. 1880).
38. *Essays and Notes,* p. 193.
39. Jerome Hall, *General Principles of Criminal Law* (Indianapolis: Bobbs-Merrill, 1947), p. 181. See also footnote 45, in which Professor Hall states what all students of Holmes should never forget, namely, that the greatest error regarding that American Olympian is to treat him as a systematic philosopher.
40. Hall, p. 48.
41. Hall, p. 176.
42. Hall, p. 178.
43. Hall, pp. 179–180. Cf. Jerome Hall, *Theft, Law and Society* (Boston, 1935), p. 292, n. 12: "Many primitive peoples never punish children."
44. Beccaria, *Dei Delitti e della pene* (1764), I, 215ff.
45. *Essays and Notes,* p. 189. Note to *U. S. v. Anthony* (1873), 11 Blatchford 200. On *ignorantia legis,* see Hall, *General Principles,* ch. 11. Professor Hall has also shown that the reform attitude to the criminal law expressed by Holmes in "The Path of the Law" *American Law Review,* X, 470 (1897), is absent from his *Common Law* (Boston, 1881). Holmes was more positivistic than Green in the 1870's.
46. *Dissenting and other Opinions of Justice Holmes* (1929 ed.), pp. 303, 307, 311.
47. *Essays and Notes,* p. 198. Also p. 146: "A frequent cause of perplexity in law is the loose way in which legal terms are used, the same term being used to express different things. The term 'assault' is one of these terms."
48. *Essays and Notes,* p. 199. Typical of Green is his insistence *after* an historical exegesis of the laws of slander and libel that "the latest decided cases on this subject *make the law*" (*ibid.,* p. 70). This is from the essay in *American Law Review,* VI, 593–633 (July 1872), which elicited Holmes's admiration (cf. text citation to note 27 above).
49. *Essays and Notes,* p. 204.
50. *Essays and Notes,* p. 209.
51. *Essays and Notes,* pp. 157–58. "The certainty of special pleading comes from the fact that it was based on scholastic logic," whereas on the Continent, the Roman law was available before the age of scholasticism (p. 158).
52. *Essays and Notes,* p. 168. This ideo-motor theory looms large in James's *Psychology* and in Fouillée's *Evolutionnisme des idées-forces* (1890). In England as early as 1866 the two influential psychologists, Bain and Lewes, were discussing the ideo-motor theory, as the following entry of Lewes' diary indicates: "June 3, 1866: Kant . . . After dinner Bain called and stayed till eleven discussing philosophical questions. Together

we groped our way to some explanation of the organic differences between the receptive and active intellects. He began by remarking how men of active productivity were almost always men of small receptivity, and too impatient of *doing*, were almost always incapable of learning. I suggested that in them the reflex was more *direct* than in the receptive natures, an idea or emotion rapidly discharging itself in a result of action, instead of exercising a wide reflex on the sensibilities and awakening a complex ideal precursor to the act" (Anna T. Kitchel, *George Henry Lewes and George Eliot, A Review of Records* (New York: John Day, 1933), p. 238). We have already noted the use of Bain as a text in the university lectures at Harvard by Wright and Green.

53. *Essays and Notes*, p. 169.

54. *American Bar Association Reports* (1896), XIX, 319–342.

55. *Essays and Notes*, p. 205.

56. *Pensées de Nicole*, "Des abus de la prévention."

57. *C. P.*, 5.12. Professor Frederick Green's testimony of his father's relations to Holmes, Wright, James, Peirce, and Warner, is contained in his letters to me.

58. "Melville M. Bigelow," by Brooks Adams, *Boston University Law Review*, I, 168 (1921).

59. Chauncey Wright, "McCosh on Intuitions," *Nation*, I, 627 (1865); "German Darwinism," *Nation*, XXI, 168 (1875).

60. *Nation*, XXI, 225–26 (1875).

61. O. W. Holmes, Jr., "Ideals and Doubts," *Illinois Law Review*, vol. X (1915), reprinted in *Collected Legal Papers*, p. 303.

62. O. W. Holmes, Jr., Introduction to *The Rational Basis of Legal Institutions* (1923), reprinted in Max Lerner's *The Faith and Mind of Justice Holmes* (Boston, 1943), pp. 399–400.

VIII. EVOLUTIONARY PRAGMATISM IN HOLMES'S THEORY OF THE LAW

1. Holmes's undergraduate essay on "Plato" was printed in *The University Quarterly*, IV, 205 (Oct. 1860).

2. Letter of Holmes to Cohen, February 5, 1919, in "The Holmes-Cohen Correspondence," edited by F. S. Cohen, *JHI*, IX, 14–15 (Jan. 1948).

3. O. W. Holmes, *Medical Essays 1842–1882* (5th ed.; Boston, 1888), p. 219; lecture iv, "Border Lines of Knowledge in Some Provinces of Medical Science," delivered November 6, 1861. Pages 103–172: lecture ii, "The Contagiousness of Puerperal Fever," was first printed in 1843, an independent and notable anticipation of Semmelweis' and Pasteur's investigations of bedside fever.

4. Mark DeWolfe Howe, *Touched with Fire: Civil War Letters and Diary of Oliver Wendell Holmes, Jr., 1861–1864* (Cambridge, Mass., 1946), pp. 95–97.

5. I am indebted to Professor Mark DeWolfe Howe for the privilege

of reading these recently discovered diaries of 1866 and 1867. Cf. his
"O. W. Holmes, Jr., Counsellor at-Law," *Publications of Brandeis Lawyer's
Society* (Oct. 1947). For Holmes's letters to James, see Perry, *Thought and
Character*, I, 505ff.; II, 457ff.

6. Letter of Holmes to Pollock, August 30, 1929, in *Holmes-Pollock
Letters*, II, 252. Cf. Holmes's letter to Max C. Otto, September 26, 1929 in
Journal of Philosophy, XXXVIII, 391 (1941); also the letter to Morris R.
Cohen, September 14, 1923, in *JHI*, IX, 34–35 (1948): "That we could
not assert necessity of the order of the universe I learned to believe from
Chauncey Wright long ago. I suspect C. S. P[eirce] got it from the same
source."

7. Quoted by Chauncey Wright in his note on Maine in the *Nation*,
XX, 411 (June 17, 1875).

8. *Ibid.*, see also the next issue of the *Nation*, vol. XXI (July 1,
1875) for another note by Wright on Maine, and Wright's letter to Grace
Norton, July 10, 1875, in *C. W.*, p. 340.

9. *C. W.*, p. 351. See also, Chauncey Wright's pragmatic ideas of
jurisprudence in *Philosophical Discussions*, pp. 258f., noted by Max H.
Fisch, "One Hundred Years of American Philosophy," *Philosophical Review*, LVI, 368–69 (July 1947).

10. Holmes, *Common Law*, p. 1. This book had a thirty-seventh printing in 1945.

11. *Justice Holmes to Dr. Wu: An Intimate Correspondence 1921–
1932* (New York, n. d.), p. 31; letter to Dr. Wu, July 21, 1925.

12. *Lochner* v. *New York*, 198 U. S. 45, 75–76 (1905), Holmes *J.*
dissenting.

13. *Dissenting Opinions of Justice Holmes*, p. 50, quoted from 250
U. S. 624.

14. *Common Law*, p. 1.

15. *Common Law*, pp. 2, 36.

16. *Common Law*, p. 14.

17. *Common Law*, pp. 15, 16.

18. *Common Law*, pp. 35f.

19. *Common Law*, p. 35.

20. *Common Law*, p. 75. See also, *Adkins* v. *Children's Hospital*, 261
U. S. 525 (1923), Holmes, *J.* dissenting: "The criterion of constitutionality
is not whether we believe the law to be for the public good."

21. *Noble State Bank* v. *Haskell*, 219 U. S. 110 (1911). Cf. Holmes's
dictum that law is "a statement of the circumstances/in which the public
force will be brought to bear upon men through the courts"; *American
Banana Co.* v. *United Fruit Co.*, 213 U. S. 347, 356.

22. *Common Law*, p. 41.

23. *Common Law*, p. 108.

24. *Common Law*, pp. 57, 134, 139; cf. J. Bentham, *The Limits of
Jurisprudence Defined, Being Part Two of An Introduction to the Principles
of Morals and Legislation*, edited by C. W. Everett (New York, 1945), p.

292n., and ch. 12 ("Consequences") in Bentham's *Introduction*, part one.

25. *Common Law*, pp. 162f.

26. *Common Law*, p. 163.

27. *Albany Law Journal*, XXVI, 444–446 (Dec. 16, 1882). Cf. Joseph B. Warner's review of Holmes's *Common Law* in *American Law Review*, XV, 331–338 (May 1881): "Holmes has the genius fit for this legal embryology . . . the historical imagination . . . the eye for the growth of an idea which belongs to the historian of institutions" (p. 332).

28. *Common Law*, pp. 164–340 (lectures vii–ix: Contract I. History, II. Elements, III. Void and Voidable Contracts).

29. *Albany Law Journal*, XXVI, 444–446 (Dec. 16, 1882).

30. O. W. Holmes, Jr., "The Path of the Law," *Harvard Law Review*, X, 456f (1896–97).

31. *Ibid.* Cf. O. W. Holmes, Jr. *American Law Review*, VI, 724 (1872): "The only question for the lawyer is, how will the judges act? Any motive for their action, be it constitution, statute, custom or precedent, which can be relied upon as likely *in the generality of cases to prevail,* is worthy of consideration as one of the sources of law, in a treatise on jurisprudence."

32. *Justice Holmes to Dr. Wu*, p. 37; letter of August 26, 1926. For a similar reference to military conscription as evidence of the way considerations of public safety prevail over the individual's will, see *Common Law*, p. 43.

33. Quoted by Bent, *Oliver Wendell Holmes*, p. 9.

34. *Common Law*, p. 43.

35. *Southern Pacific* v. *Jensen*, 244 U. S. 205 (1917), Holmes J. dissenting.

36. Holmes, *Speeches*, p. 18.

37. *Common Law*, p. 41.

38. *Common Law*, p. 213.

39. *Ibid.* Holmes gives a footnote reference to Wake's *Evolution of Morality* on the instinct of possession.

40. "Views of public policy are taught by the interests of life. These interests are fields of battle"; O. W. Holmes, Jr., "Privilege, Malice, and Intent," *Harvard Law Review*, VIII, 7 (1894). Another Darwinian comment occurs in Holmes's discussion of the law of sales. A breach of warranty of *quality* makes a sale voidable but a different kind of goods from what was originally offered may make a sale voidable. Holmes says, "When does a difference in quality give rise to a difference in kind? It is a question for Mr. Darwin to answer"; O. W. Holmes, Jr., "The Theory of Torts," *American Law Review*, VII, 652 (1872–73). Chauncey Wright had at this time made a similar logical point in his defense of Darwin's theory of species merging insensibly into one another by accumulating variations until a large difference of degree became a taxonomic difference; cf. his review of Mivart's *Genesis of Species* in the *North American Review* (1872).

41. Letter of Holmes to Cohen, March 13, 1925, *JHI*, IX, 44 (Jan. 1948).

42. Cf. John Dewey, "Logical Method and the Law," *Cornell Law Quarterly*, X, 17–27 (1924). Morton G. White, "The Revolt Against Formalism in American Social Thought of the Twentieth Century," *JHI*, VIII, 135ff (1947), especially footnote 24.

43. Letter of Holmes to Cohen, July 21, 1920, *JHI*, IX, 19 (Jan. 1948).

44. Letter of Cohen to Holmes, August 7, 1920, *JHI*, IX, 20 (Jan. 1948).

45. *Justice Holmes to Dr. Wu*, p. 40; letter of December 5, 1926.

46. *Justice Holmes to Dr. Wu*, p. 40; letters of January 30, 1928, and July 9, 1928.

47. Letter of Holmes to Cohen, September 6, 1920, *JHI*, IX, 23 (Jan. 1948).

48. Letter of Holmes to Cohen, August 10, 1922, *JHI*, IX, 32 (Jan. 1948).

49. Letter of Holmes to Otto, September 26, 1929, *Journal of Philosophy*, XXXVIII, 391 (1941). Cf. Holmes's earlier use of his "can't help" definition of truth in his criticism of the "natural law" philosophy; "Ideals and Doubts," *Illinois Law Review*, vol. X (1915): "When I say that a thing is true, I mean that I cannot help believing it. I am stating an experience as to which there is choice. But there are many things that I cannot help doing that the universe can, I do not venture to assume that my inabilities in the way of thought are inabilities of the universe. I therefore define the truth as the system of my limitations, and leave absolute truth for those who are better equipped. With absolute truth I leave absolute ideal of conduct equally on one side . . . I used to say, when I was young, that truth was the majority vote of that nation that could lick all the others."

50. Holmes, *Speeches* (1918), p. 17 ("The Law," Feb. 5, 1885). In an unpublished letter, Holmes, past ninety, wrote: "I was rather shoved than went into the law when I hankered for philosophy. I am glad now, and even then I had a guess that perhaps one got more from philosophy on the quarter than dead astern"; quoted by Felix Frankfurter, article on Holmes, *Dictionary of American Biography*, XXI, suppl. 1 (New York, 1944), p. 419, who also notes that Holmes "came to maturity when Darwin began to disturb ancient beliefs. If Genesis had to be 'reinterpreted' no texts of the law, however authoritative, could claim sanctity . . . In his formative years he found most congenial the company of speculative minds like William James and Charles S. Peirce and Chauncey Wright" (p. 426). Cf. Holmes's college autobiography in *New England Quarterly* (March 1933).

IX. THE PHILOSOPHICAL LEGACY OF THE FOUNDERS OF PRAGMATISM

1. *C. P.*, 6.498f.
2. Lovejoy, *Great Chain of Being*, ch. ix.

3. See Morton G. White, *The Origin of Dewey's Instrumentalism* (New York, 1943); also his "The Revolt Against Formalism in American Social Thought of the Twentieth Century," *JHI*, VIII, 131–152 (1947).

4. For a more thorough discussion of a pragmatic relativism that is not subjective, see Addison W. Moore, *Pragmatism and Its Critics* (Chicago, 1910), and C. I. Lewis, *Analysis of Knowledge and Valuation*, The Carus Lectures (LaSalle, Ill., 1946). In this work Professor Lewis acknowledges his intellectual debt to Peirce and Dewey, and in an earlier important work, *Mind and the World Order* (New York, 1929), he calls his philosophy "conceptual pragmatism." A history of pragmatism in the twentieth century would have to do justice to the varieties of pragmatism in C. I. Lewis, G. H. Mead, Addison W. Moore, H. Heath Bawden, Boyd H. Bode, Max C. Otto, Harry A. Overstreet, M. R. Cohen, J. H. Randall, Jr., Irwin Edman, Percy W. Bridgman, Ernest Nagel, Sidney Hook, Hans Reichenbach, Charles W. Morris, F. S. C. Schiller, Giovanni Papini, Henri Bergson, Edouard LeRoy, Henri Poincaré, Giovanni Vailati, and others, outside of professional philosophy, whose common pragmatic tendencies are discernible beneath a rich diversity of teachings even when they are in sharp conflict on specific questions.

Index

Pennsylvania Paperbacks

Pennsylvania Paperbacks continued